# Data-Driven Storytelling

# A K PETERS VISUALIZATION SERIES

Series Editor: Tamara Munzner

### Data-Driven Storytelling
*Nathalie Henry Riche, Christophe Hurter, Nicholas Diakopoulos,*
*and Sheelagh Carpendale*
2018

### Information Theory Tools for Visualization
*Min Chen, Miquel Feixas, Ivan Viola, Anton Bardera, Han-Wei Shen,*
*and Mateu Sbert*
2016

### Visualization Analysis and Design
*Tamara Munzner*
2014

# Data-Driven Storytelling

Edited by
Nathalie Henry Riche, Christophe Hurter,
Nicholas Diakopoulos, and Sheelagh Carpendale

CRC Press
Taylor & Francis Group
Boca Raton London New York

CRC Press is an imprint of the
Taylor & Francis Group, an **informa** business

AN A K PETERS BOOK

CRC Press
Taylor & Francis Group
6000 Broken Sound Parkway NW, Suite 300
Boca Raton, FL 33487-2742

International Standard Book Number-13: 978-1-138-48225-8 (Hardback)
International Standard Book Number-13: 978-1-138-19710-7 (Paperback)

### Library of Congress Cataloging-in-Publication Data

Names: Riche, Nathalie Henry, editor.
Title: Data-driven storytelling / [contributions by] Nathalie Henry Riche, Christophe Hurter, Nicholas Diakopoulos, Sheelagh Carpendale.
Description: Boca Raton, Florida : Taylor & Francis/CRC Press, [2018] | Includes bibliographical references.
Identifiers: LCCN 2017060385 | ISBN 9781138197107 (pbk. : alk. paper) | ISBN 9781138482258 (hardback : alk. paper) | ISBN 9781315281575 (ebook)
Subjects: LCSH: Narration (Rhetoric) | Digital storytelling. | Visual communication. | Storytelling in mass media.
Classification: LCC P96.N35 D38 2018 | DDC 808/.036--dc23
LC record available at https://lccn.loc.gov/2017060385

**Visit the Taylor & Francis Web site at**
**http://www.taylorandfrancis.com**

**and the CRC Press Web site at**
**http://www.crcpress.com**

# Contents

# Acknowledgments

The editors would like to thank **Schloss Dagstuhl—Leibniz Center for Informatics** for making it possible to gather an interdisciplinary group of renowned data visualization researchers, practitioners, and data journalists at the Data-Driven Storytelling seminar (16061), and **Schloss Dagstuhl—NSF Support Grant** for sponsoring the participation of outstanding junior researchers, as well as **Microsoft** for sponsoring the participation of data journalists at the event. This Dagstuhl seminar was the catalyst for this book.

Also, they would like to thank **Kennedy Elliot**, a developer and journalist who reports, designs, and develops interactive graphics and data visualizations for the news and has previously worked at The Washington Post, the Guardian US, and the Associated Press, and **Stefan Wehmeyer**, a civic technologist and data activist from Berlin, Germany, who founded the Freedom of Information portal FragDenStaat.de and worked as a data journalist for the non-profit investigative center correctiv.org.

Kennedy and Stefan both contributed to discussions at the seminar which informed several of the chapters.

# Editors

**Nathalie Henry Riche** is a researcher at Microsoft Research in the EPIC (Extended Perception Interaction Cognition) group led by Ken Hinckley and part of the broader HCI@MSR effort. She received a joint PhD in computer science from the University of Paris-Sud/Inria, France and the University of Sydney, Australia in 2008. Nathalie conducts research at the intersection of human–computer interaction (HCI) and data and information visualization. Her passion is to create novel interactive visualizations to help people explore and think with data. Her recent research focused on enabling people to communicate insights with data visualizations. She studies data-driven storytelling techniques used "in the wild" and develop authoring interfaces to craft and deliver novel genres of data-driven stories.

Nathalie would like to thank Microsoft Research for supporting this effort and the seminar leading to this book.

**Christophe Hurter** received his PhD (2010) in computer science from the Toulouse University, France. Until 2015, he was an associate professor at the Interactive computing laboratory (LII) of the French Civil Aviation University (ENAC) in Toulouse, France. Since 2016, he is a professor of computer science at the same university leading the Data Visualization group. He is also associate researcher at the research center for the French military Air Force (CReA), Salon de Provence, France. His research interests cover information visualization and HCI, particularly the visualization of multivariate data in space and time, the design of scalable visual interfaces, and the development of image-based infovis algorithms.

Christophe would like to thank the French National Agency for Research (Agence Nationale de la Recherche—ANR) under grant ANR-14-CE24-0006-01 project "TERANOVA".

**Nicholas Diakopoulos** is an assistant professor at the Northwestern University School of Communication, where he directs the Computational Journalism Lab. He is also a Tow Fellow at the Columbia University School of Journalism and an associate professor at the University of Bergen Department of Information Science and Media Studies. He received his PhD in Computer Science from the School of Interactive Computing at Georgia Tech where he co-founded the program in Computational Journalism. His research spans computational and data journalism, including topics such as algorithmic accountability reporting, algorithmic transparency, social computing and media in news contexts, and storytelling with data visualization.

Nick would like to thank the University of Bergen VisMedia project which supported efforts on this project, as well as Microsoft Research for financial assistance in attending the Dagstuhl workshop.

**Sheelagh Carpendale** is a full professor at the University of Calgary, where she holds a Canada Research Chair in information visualization and NSERC/AITF/SMART Technologies Industrial Research Chair in interactive technologies. She has many received awards, including the Canadian E.W.R. NSERC STEACIE Award and a BAFTA (British Academy of Film & Television Arts) Interactive Award. Sheelagh leads the Innovations in Visualization (InnoVis) research group and initiated the interdisciplinary graduate program, Computational Media Design. Her research draws upon her combined backgrounds in computer science and visual arts, benefiting from the interdisciplinary cross-fertilization to enable the design of innovative, people-centred information technologies. By studying how people interact with information both in work and social practices, she works toward designing more natural, accessible, and understandable interactive visual representations of data.

Sheelagh would like to thank her research support from National Science and Engineering Research Council (NSERC), Alberta Innovates Technology Futures (AITF), and SMART Technologies. She would also like to thank her collaborators and students for on-going stimulating intellectual discussions.

# Contributors

**Fereshteh Amini** is a PhD candidate at the Department of Computer Science, University of Manitoba where she has also received her MSc degree. Her work lies at the intersection of HCI and Information visualization with primary focus on design and evaluation of information visualization tools and techniques for visual communication of data to broad audiences.

**Benjamin Bach** was a postdoc at Harvard University, Monash University, as well as the Microsoft Research-Inria Joint Centre, before joining the University of Edinburgh as an assistant professor in 2017. Benjamin was visiting researcher at the University of Washington and Microsoft Research during 2015. He obtained his PhD in 2014 from the Université Paris Sud, where he worked at the Aviz Group at Inria.

**Lyn Bartram** is an associate professor in the School of Interactive Art and Technology at Simon Fraser University and director of the Vancouver Institute of Visual Analytics, engaging researchers, practitioners, and organizations with challenges and opportunities in the emerging universe of big data. Her work explores the intersecting potential of interactive information technologies, ubiquitous computing and computational media from both theoretical and applied perspectives, most recently with a focus on personal visualization and sustainable cities. She holds a BA in political science from UBC, an MMath from University of Waterloo, and a PhD in computer science from Simon Fraser University. She is a member of IEEE, ACM, and the Pacific Institute of Climate Solutions.

**Dominikus Baur** is a data visualization expert and interaction designer from Germany. In his PhD on visualizing music listening histories, he explored everyday storytelling and reminiscing through personal data. As a consultant he has helped clients such as Google, the OECD, and Microsoft Research. He is a regular speaker at conferences and a trainer on data visualization.

**Gordon Bolduan** is a computer scientist at Saarland University and journalist. He is interested in the complete process of data-driven storytelling, from exploring data to visualizing it. He is responsible for science communication at the Cluster of Excellence "Multimodal Computing and Interaction" and at the Research Center for IT Security, Privacy and Accountability (CISPA). Prior to this he worked as reporter for the magazine, *Technology Review*.

Gordon thanks Dr. Roswitha Bardohl, Schloss Dagstuhl—Leibniz Center for Informatics.

**Jeremy Boy** is the data visualization and design specialist at UN Global Pulse. His research focuses primarily on the learning processes people undergo when confronted with data-rich interactive media. Jeremy has a PhD in information and communication sciences, and a masters degree in graphic/multimedia design.

Jeremy thanks Schloss Dagstuhl and NSF for the Support Grant for Junior Researchers to travel to the Dagstuhl seminar. He also thanks the organizers of the seminar/editors of the book for making this possible.

**Matthew Brehmer** is a post-doctoral researcher at Microsoft who specializes in data visualization for storytelling and journalism. He completed his PhD in computer science in 2016 at the University of British Columbia, where he was a member of Tamara Munzner's InfoVis group. He completed his MSc in HCI in 2011 (also at UBC) and BComp in cognitive science from Queen's University in 2009.

**Fanny Chevalier** is an assistant professor at University of Toronto, Canada, whose research focuses on interactive tools supporting visual analytics, visualization education, and creative activities. She is interested in exploring how data stories can most effectively reach out to the youth and general public, help citizens better understand news information sources, and foster critical thinking.

**Paolo Ciuccarelli** is head of the BSc and MSc in communication design and member of the Design PhD board, both at Politecnico di Milano; Paolo has been awarded the 2017 Fellowship at ISI Foundation. Coeditor of the journal "Big Data and Society" (SAGE)—demo section—he founded the DensityDesign Research Lab where he holds the position of scientific director. He is member of the Steering Committee of the "Reassembling the Republic of Letters"—a Digital Humanities COST Action where he leads the Visualization and Communication Working Group.

**Marian Dörk** is a research professor for information visualization at the Urban Futures Institute for Applied Research of the University of Applied Sciences Potsdam, where he codirects the Urban Complexity Lab, a transdisciplinary research space at the interface between computing, design, and culture. His research studies the potential of visualization to support exploratory information practices.

**Steven M. Drucker** is a principal researcher at Microsoft Research focusing on HCI with data. He is also an affiliate professor at UW's CSE Department. He has demonstrated his work on stage with Bill Gates and Satya Nadella, shipped data visualization products, was written up in the *New York Times*, filed over 108 patents, and published papers on technologies as diverse as information visualization and spectator-oriented gaming.

**Tim Dwyer** completed his PhD at the University of Sydney in 2005, and then was a postdoctoral research fellow at Monash University until 2008. He moved to Microsoft in 2008, first as a researcher and then an engineer with Visual Studio. In 2012, he returned to Monash University where he now codirects the Immersive Analytics Initiative which explores the role of emerging display and interaction technologies for understanding and communicating about complex data.

**Jason Dykes** is a professor in visualization at City, University of London. He uses his background in GIScience and cartography to inform his InfoVis research and practice. Maps allow us to link places and features in multiple ways as we read them—perhaps following linear features, perhaps cutting across them, but always making links. Can we use maps to develop meaningful narrative, to guide discovery, and to inform data-driven stories?

Many of the ideas and perspectives that informed Jason's contributions to the Dagstuhl seminar and subsequent chapter were informed by discussions over the years with giCentre colleagues.

**Christina Elmer** leads the data journalism team at Spiegel Online in Hamburg, Germany. She is a member of the editorial board, and also trains journalists in data reporting and online research. Before she joined Spiegel Online in 2013, Elmer worked at *Stern* Magazine's investigative unit. Her journalistic career began in 2007 at the German press agency dpa, where she was part of a team which set up Germany's first department for data journalism and computer-assisted reporting. Elmer studied journalism and biology. As a board member of netzwerk recherche, Germany's largest association supporting investigative reporters, she is actively engaged in pushing data journalism forward and encouraging colleagues to use new techniques and tools.

**Yuri Engelhardt** is an assistant professor at the University of Amsterdam. He loves thinking about the universal principles of representing information visually. His academic path has included an MSc in medicine, a PhD in computer science, studying cognitive psychology, teaching media studies, philosophy and design at a faculty of humanities and at an art school, and supervising PhD students in geovisual analytics. Yuri is passionate about sustainable human well-being.

Yuri is grateful for all the nurturing support from Lieve Witteveen—the most amazing person on this planet.

**Theresia Gschwandtner** is a senior researcher at TU Wien and deputy research chair of the Centre of Visual Analytics Science and Technology. Her main research interests include visual analytics and information visualization of data quality, uncertainty, and healthcare, as well as guidance in visual analytics.

Theresia thanks the Centre for Visual Analytics Science and Technology CVAST, funded by the Austrian Federal Ministry of Science, Research, and Economy in the exceptional Laura Bassi Centres of Excellence initiative (822746).

**Jessica Hullman** is an assistant professor in the Information School at the University of Washington. Her research focuses on how visualizations and other interactive representations can be used to make complex information more comprehensible to broad audiences at scale. Her work often starts by identifying principles from experts' visualizations and then instantiating these in automated tools.

**Samuel Huron** is a designer and computer scientist. He is an associate professor at Telecom Paris Tech and associate director of the CoDesignLab. His research interests focus on the design process of visual representation, design methods apply to research, and visualization for broad audiences. His research is grounded in 15 years of experience in the web and communication industry, where he worked for a broad range of civic, cultural, and corporate clients.

**Ulrike Köppen** is responsible for BR Data, the data and interactive team of the public broadcaster Bavarian Broadcasting/ARD. The interdisciplinary team is dedicated to all forms of data journalism and interactive storytelling. Köppen is passionate about combining data, storytelling and journalistic research to produce new forms of journalism. Together with colleagues she founded ddjmonaco, a regular tech meeting for journos, coders, and open data people in Munich. She studied comparative literature and semiotics in Munich and Paris, and worked at arte, Süddeutsche Zeitung and the news agency dapd before building up a web innovation team at Bavarian Broadcasting.

**Robert Kosara** is a research scientist at Tableau Software. His focus is on the communication of data through visualization and visual storytelling. Understanding what makes story effective is the first step toward better tools that can help anybody use data to get their points across and support better decisions.

**Bongshin Lee** is a senior researcher at Microsoft Research. She explores innovative ways for people to create visualizations, interact with their data, and share data-driven stories. She has been recently focusing on helping people explore the data about themselves and share meaningful insights with others by leveraging visualizations. She holds MS and PhD degrees in computer science from University of Maryland at College Park.

**Giuseppe Santucci** is an associate professor at the Department of Computer Science of Sapienza Università di Roma. His main research activities concern HCI, information visualization, focusing on quality aspects of information visualization techniques, and the application of visual analytics on predictive scenarios.

**Jonathan Schwabish** of the Urban Institute is an economist, writer, teacher, and creator of policy-relevant data visualizations. He is considered a leading voice for clarity and accessibility in how researchers communicate their findings. He helps nonprofits, research institutions, and governments at all levels and improves how they communicate their work and findings to their partners, constituents, and citizens.

Jonathan thanks the Urban Institute.

**John Stasko** is an internationally recognized professor at Georgia Institute of Technology in the areas of information visualization and visual analytics. His work develops ways to help people visually explore, analyze, and make sense of data in order to solve problems. Stasko, an IEEE Fellow and ACM CHI Academy member, received the IEEE VGTC Visualization Technical Achievement Award in 2012.

**Moritz Stefaner** works as a "truth and beauty operator" on the crossroads of data visualization, information aesthetics, and user interface design. With a background in cognitive science (BSc, University of Osnabrueck) and interface design (MA, University of Applied Sciences Potsdam), his work beautifully balances analytical and aesthetic aspects in mapping abstract and complex phenomena. http://truth-and-beauty.net, http://datastori.es.

**Charles Stolper** is an assistant professor in the Mathematics and Computer Science Department at Southwestern University in Georgetown, TX. He received his PhD from the School of Interactive Computing at the Georgia Institute of Technology in Atlanta, GA. He conducts research on information visualization and data science topics, with a focus on languages for graph visualization, exploration of sport data, progressive visual analytics, and data storytelling.

**Alice Thudt** is a PhD candidate in computational media design in the InnoVis Group at the University of Calgary. Her research explores how visualizations of personal data can support autobiographical storytelling and self-reflection. Recent projects have investigated the role of subjectivity in personal visual narratives, design of visualization techniques to reflect autobiographical memories, as well as the creation of visual mementos.

Alice thanks the organizers and participants of the Dagstuhl seminar on data-driven storytelling for their input and the funding agencies AITF, NSERC, and Smart Technologies for making this research possible.

**Melanie Tory** is a senior research scientist at Tableau Research. Her research explores techniques and tools to support interactive visual data analysis, including sensemaking, collaboration, and storytelling. Before joining Tableau, Melanie was an associate professor in visualization at the University of Victoria. She earned her PhD in computer science from Simon Fraser University and BSc from the University of British Columbia. She is associate editor of IEEE Computer Graphics and Applications and has served as paper cochair for the IEEE Information Visualization and ACM Interactive Surfaces and Spaces conferences.

**Barbara Tversky** is a professor emerita of psychology at Stanford University and a professor of psychology and education at Teachers College, Columbia University. A cognitive psychologist, she has cognitive psychologist, she has worked on memory, categorization, spatial thinking, language, event perception and cognition, storytelling, diagrammatic reasoning, sketching, gesture, creativity, and their applications. She has enjoyed collaborations with linguists, philosophers, neuroscientists, biologists, geographers, chemists, computer scientists, engineers, architects, museum educators, artists, and more.

Barbara expresses gratitude to Nathalie Riche and Christophe Hurter for their insightful suggestions and to grants: NSF CHS 1513841, HHC 0905417, IIS-0725223, IIS-0855995, and REC 0440103 & John Templeton Foundation Varieties of Understanding project.

**Xaquín González Veira** is an interactive, visual journalist. Most recently, he led *The Guardian*'s 40-people visuals team—an experiment of graphics, interactive, multimedia, and picture editors. Before that, he worked at National Geographic, The *New York Times*, *Newsweek*, *El Mundo*, and *La Voz de Galicia*. His latest obsession? Emotional connections in data-driven visual storytelling.

Xaquín thanks Microsoft for allowing him to meet the most inspiring collection of talent in this seminar, his dad Xoán G. (his first dataviz mentor), Cris and Roi for "forcing" him to come to Dagstuhl.

**Jagoda Walny** recently completed his PhD in computer science in the InnoVis Group at the University of Calgary's Interactions Lab. She researches how to create new visualization interfaces by observing thinking sketches, that is, the visuals that people create day-to-day to help themselves to think in individualized ways.

Jagoda thanks NSERC and SMART Technologies for providing travel funding.

**Benjamin Wiederkehr** is an interaction designer at Interactive Things in Zurich, Switzerland, with a focus on information visualization and interface design. He explores opportunities to innovate through the combination of design and technology, to simplify complex data in order to raise awareness, as well as to tell stories with an open intent and meaningful impact.

**Jarke van Wijk** is a full professor in visualization at the Department of Mathematics and Computer Science of Eindhoven University of Technology. His main research interests are information visualization and visual analytics, aiming to develop new methods and techniques that enable people to understand and explore large and complex datasets. He received multiple awards, and the research of his group has led to two successful start-up companies.

**Jo Wood** is a professor of visual analytics at the giCentre, City, University of London, where he designs, builds and applies data visualization software to analyze patterns in complex datasets. He has interests in visualization and analysis of sustainable transport and narrative in visual analytic design.

# CHAPTER 1

# Introduction

## Nathalie Henry Riche

*Microsoft Research*

## Christophe Hurter

*French Civil Aviation University (ENAC)*

## Nicholas Diakopoulos

*Northwestern University*

## Sheelagh Carpendale

*University of Calgary*

## CONTENTS

Today, data visualizations are everywhere. They form a significant and often integral part of contemporary media. Stories supported by facts extracted from data analysis proliferate in many different ways in our analog and digital environments including printed infographics in magazines, animated images shared on social media, and interactive online visualizations tightly integrated with news stories on leading media outlets. However, while data visualization may be at the heart of data-driven stories, the concepts are not synonymous. We explore these distinctions in this book.

The appearance of several landmark books such as Bertin's *Semiology of Graphics: Diagrams, Networks, Maps* (original French 1967, English translation 1983), Tukey's *Exploratory Data Analysis* (1977), Tufte's *The Visual Display of Quantitative Information* (1984), and Cleveland and McGill's *Dynamic Graphics for Statistics* (1988) set the stage for the emergence of a recognized area of research in visualization by the late 1980s. This emergence was fueled by the increasingly prevalent possibility of using computers to make data visual and interactive and was inspired by the popular idiom that "a picture is worth a thousand words." While the research community has favored phrases such as "scientific visualization," "knowledge visualization," and "information visualization," in general, media tends to use the more encompassing phrase "data visualization." Due to the importance of visualization to data-driven stories in journalism, we also use the phrase "data visualization."

Data has been represented visually from the early history of humans. Perhaps the oldest examples are those documented by Marshack, where he shows examples of keeping records through scratches on surfaces such as Ishango bones (18,000–20,000 BC) and Lebombo bones (35,000 BC) (Marshack 1991). While these considerably predate written language, humans were clearly using visual representations to help themselves understand their world. If the ability to make scratches on bones is considered the first technology boon to data visualization, large advances in what is possible to achieve with data visualization can be associated with the availability of developing technologies. In broad steps, this includes the use of clay tablets in Mesopotamia, the development of paper in Egypt, Johannes Gutenberg's printing press in 1440, Konrad Zuse's computer in 1940, and most recently in the late 1980s when Tim Berners-Lee made the Internet widely accessible via the World Wide Web. In terms of our focus on data-driven stories, this last factor of making computational power and prowess widespread has been crucial. It is this factor that has made it possible for all types of media to consider incorporating evidence, portrayed by the visualization of the data that supports a given story, directly into the presentation of that story.

We think this movement towards data-driven stories, which is apparent in both the data visualization research community and the professional journalism community, has the potential to form a crucial part of keeping the public informed, a movement sometimes referred to as the democratization of data – the making of data understandable to the general public. This exciting new development in the use of data visualization in media has revealed an emerging professional community in the already complex group of disciplines involved in data visualization. Data visualization has roots in many research fields including perception, computer graphics, design, and human-computer interaction, though only recently has this expanded to include journalism.

## RESEARCH IN DATA VISUALIZATION: FROM UNDERSTANDING TO EXPLORATION TO DATA STORYTELLING

Early research in data visualization focused on producing static images and quantifying the perception of different visual encodings to understand data visually (Card, Mackinlay, and Shneiderman 1999). The vast majority of research since then has focused on designing and implementing novel interfaces and interactive techniques to enable data exploration. Major advances in visual analytics and big-data initiatives have concentrated on integrating machine learning and analysis methods with visual representations to enable powerful exploratory analysis and data mining (Thomas and Cook 2005). As interactive visualizations play an increasing role in data analysis scenarios, they also started to appear as a powerful vector for communicating information. The popularity of JavaScript Web technology and the availability of the D3 toolkit (Bostock, Ogievetsky, and Heer 2011) also made it possible for a wider range of people to create data visualizations. Being able to easily share interactive data visualizations on the Web also increased the democratization of interactive visualizations. Since the field is mature enough, it is now time to understand how these powerful interactive and dynamic data visualizations can play a role in communicating information in novel ways.

## PRACTICE IN DATA JOURNALISM: FROM COMMUNICATION TO DATA EVIDENCE TO DATA STORYTELLING

Journalism has always been about communication, finding relevant stories and disseminating them publically, observing events, gathering information, and telling this information to the general public in a manner that can be understood and is both interesting and relevant (Kovach and Rosenstiel 2007). There has been an increasing onus on the quality of this information, from the right to protect an information source, to increasing interest in documented evidence, as in photographs, audio recordings, and video recordings, all of which can be thought of as types of data. There has also been a growing consciousness that some of today's most relevant stories are buried in data. This data can be quite hard to understand in its raw formats but can become much more generally accessible when visualized. Journalists have not only begun to use standard data visualizations such as charts and maps in their stories, but are also creating new ones that are tailored to the particular data type and to the message of the story they are writing. Since journalists are now able to easily share interactive data visualizations on the Web, the democratization of data visualization is accelerating with new compelling data visualizations emerging in the media daily. This has led to extensive and practical progress on the challenges of data-driven storytelling. News sites like *The New York Times*,[*] *FiveThirtyEight*,[†] *Bloomberg*,[‡] and *The Washington Post*[§] were early movers in capturing the surge of attention and interest in consuming data-driven news by the public. By carefully

---

[*] http://www.nytimes.com/.
[†] http://fivethirtyeight.com/.
[‡] http://www.bloomberg.com/.
[§] http://www.washingtonpost.com/.

structuring the information and integrating explanation to guide the consumer, journalists help lead readers towards a valid interpretation of the underlying data. In parallel to the last section on the visualization researcher perspective, we can also say that it is now time to learn how these powerful interactive and dynamic data visualizations play a role in communicating information in novel ways.

## FORGING NEW INTERDISCIPLINARY PERSPECTIVES

At the time of formulating the possibility of this book, there was little overlap and collaboration between the two major communities involved: professional journalists who are at the forefront of making data-driven stories and academic researchers who are exploring research questions in regard to the role visualization can play in storytelling with data. This gap between research and practice has been widening as novel and innovative examples and genres of storytelling with data flourish in the media quite separately from the knowledge being built by the research community. The goal of this book is to try to close this gap by bringing together the voices of leading researchers and practitioners on data-driven storytelling. The chapter topics and their content were defined by authors with representation and participation from both communities.

Because of the rapid and practical advances in data-driven storytelling and its increasingly widespread use, we gathered several of the top practitioners from journalism and design together with visualization researchers to discuss the challenges and opportunities of data-driven communication during a Dagstuhl Seminar. Schloss Dagstuhl,* a unique venue sponsored by the German government, is a place where computer science researchers can meet to discuss currently important research questions. Dagstuhl encourages interdisciplinary discussion that includes leading thinkers from academia and industry. Founded in 1990, it has earned an international reputation as an incubator for new ideas. The four editors of this book organized the 16061 seminar† February 7th–12th, 2016. During these 5 days, a carefully selected group of 42 thinkers with diverse backgrounds: journalists, designers, perception, human-computer interaction, and visualization researchers, reflected, exchanged, and synthesized knowledge on data-driven storytelling, which led to this book.

---

* http://www.dagstuhl.de/en/.
† http://www.dagstuhl.de/en/program/calendar/semhp/?semnr=16061.

In brief, the aims of the seminar were as follows:

1. To bring together academic and industrial researchers from the human-computer interaction, cognitive psychology, information visualization, and visual analytics research communities, as well as storytelling experts from data journalism, design, art, and education.

2. To prepare a data-driven storytelling research agenda that includes a definition of data-driven storytelling, to compile examples, and to provide a detailed description of research directions in this space, and to offer a motivating list of research opportunities and challenges.

3. To investigate how the evaluation of data-driven stories can be done, including via expert critique, as well as through studies of audience comprehension, engagement, biases, and visualization literacy.

4. To discuss the ethics of data-driven storytelling authoring, identifying possible sources of bias and investigating how the lie factor of static visualizations applies to different media.

5. To compile examples of good and bad practices in application domains (data journalism, design, art, and education) and report on current processes and practices to create data-driven stories.

6. To formalize and explore the design space for novel consumption experiences in each domain. In particular, to reflect on the various advantages of different devices and input technologies (e.g., mobile phones, touch, or pen-enabled interfaces).

7. To formalize and explore the design space for novel authoring interfaces to democratize data-driven storytelling, focusing on audiences that are not able to program their own custom experiences.

8. To build individual collaborations between the seminar attendees and hence, build the community around data-driven storytelling research.

The chapter topics in this book were chosen through full group discussions. Once the topics were selected, smaller subgroups with particular interest and expertise in a given topic discussed the topic in depth. Thoughts presented in this book are the results of these conversations that were initiated at the seminar and were pursued over a year and finalized as chapters in this book.

## NOTES ON TERMINOLOGY

Since this book has emerged from cross-discipline discussions, it seems appropriate to define, for use in this book, common terms in use in all communities with often slightly variant definitions. We include the definitions of key concepts for data-driven stories below.

> **InfoVis:** the definition from Card, Mackinlay, and Shneiderman (1998) as "the use of computer-supported interactive visual representations of abstract data to amplify cognition" still serves well and provides us with our working definition of data visualization. In other words, we see data visualization as a more inclusive term in that the term "data" covers all visualization variations (scientific vis, knowledge vis, information vis), which is important because data-driven stories do not limit themselves to any particular type of data. Thus for this book, we define the following terms:
>
> > **Data visualization:** "The use of interactive, dynamic, and responsive visual representations of data to amplify cognition." However, this definition requires some unpacking.
> >
> > **Interactive:** "Mutually or reciprocally active." For example, interactive play between children is when two or more children are acting and responding to each other's acts. Interactive computer systems then are those where a person acts and the computer system responds to his and her actions and vice versa.
> >
> > **Dynamic:** "Often continuous actions or changes." On a computer, many sorts of dynamics are possible such as animations, replays, stop motion, etc., that while possibly being very useful in telling a story, may not be actually interactive.
> >
> > **Responsive:** "An action that is triggered by a previous act." The previous act could be an action in the narrative or an action taken by a person.
>
> **Data:** "Gathered, collected, modeled and produced details, calculations, and measurements, often assumed as facts, and forming the basis of reasoning, analysis, and understanding."

**Abstract data**: We unpack this term from Card et al.'s definition as explaining why we omit it in our definition of "data visualization." Some data is concrete as in coordinates like latitude and longitude, which indicate a definitive place. Other data is less concrete in that it is derived from concrete data, for example, speed is derived from time and distance. On a continuum, data can become less and less concrete as it incorporates ideas such as love and hate and expressions of these ideas, such as poetry. While InfoVis is a research discipline that focuses on the challenges of visualizing abstract data, in terms of data-driven stories, we need no such distinction. A data-driven story can be based on any type of data that has emerged from any place on the data continuum from concrete to abstract.

**Amplify cognition**: To amplify cognition we must in some manner enhance the process of cognition, which is defined as "the mental action or process of acquiring knowledge and understanding through thought, experience, and the senses." Visualizations can enhance this by assisting memory (providing externalization of complex factors) and by easing comprehension (e.g., by creating representations that appropriately leverage perception).

**Narrative**: The definition according to the *Oxford English Dictionary* is "an account of a series of events, facts, etc., given in order and with the establishing of connections between them." Narratives are a sequence of events in a set order.

**Stories**: Stories are based on a set of events from a narrative, but they may adjust the presentation by changing the order, shortening the length, adding extra context, etc. While a story is based on a narrative, it is an expression of the creative prerogative to provide interest, emphasis, etc. Stories have been used since ancient times to transfer knowledge and information over time, e.g., Homer, religious parables, and folktales. Stories package information to aid memory and recall by embedding information into characters, settings, relationships, and events.

**Data-driven stories (DDSs)**: These are stories that are data-driven in that they start from a narrative that either is based on or contains data and incorporates this data evidence, often portrayed by data graphics, data visualizations, or data dynamics, to confirm or augment a given story. A DDS often incorporates the visual data representations directly into the presentation of the story. A DDS can enhance a narrative with capabilities to walk through visual insights, to clarify and inform, and to provide context to visually salient differences.

## AUDIENCE OF THIS BOOK

This book introduces key concepts on data-driven storytelling bridging several domains and pointing to seminal research and exemplary stories from practitioners. Every chapter provides a wealth of examples of DDSs along with different analyses that are discussed from different perspectives. This discourse can serve as an initial corpus for the rapidly expanding practice around DDSs. Thus, it offers an important resource for students at all levels but particularly for advanced undergraduate and graduate students in computer science, design, journalism, and communication.

Each chapter highlights challenges and research opportunities for data-driven storytelling, synthesizing knowledge from practitioners and leading researchers in the field. We envision that it can become a seminal book for academic and industrial researchers who aim to push the state of the art in data visualization.

As data-driven storytelling is compelling in a wide range of scenarios, we envision that practitioners and data enthusiasts from many domains will find inspiration, knowledge, and practical pointers for data-driven storytelling. For example, DDSs are compelling in enterprise scenarios where the output of data analysis (often reports and slide-based presentations) and business intelligence has to be conveyed to decision makers. In education scenarios, DDSs can be an effective medium to explain complex concepts or to illustrate biological or physical mechanisms. Finally, in many other scenarios, from journalism to art and entertainment, DDSs can help communicate complex findings to a broad audience in an engaging manner.

Data-story creators may use this book as a handbook to guide the design and evaluation of the DDSs that they create. Discussions in the chapters are grounded in real-world examples from leading news media outlets and companies, and incorporate discussions from the authors of these pieces on the design process and experience, providing valuable knowledge for the practice of crafting DDSs. The chapters also point to many fundamental research findings and open new research questions that can serve to inform future DDS research including such factors as leveraging design and discovering appropriate strategies for evaluation.

## BOOK STRUCTURE

The authors in this book have a wide range of backgrounds as researchers and professionals in the fields of data visualization, design, perception and human-computer interaction, and data journalism. This diversity offers different and rich perspectives on the topic, and is revealed via the multiple facets written in each chapter. This book covers topics on perceptual and cognitive foundations of data-driven storytelling (see Chapter 2), the content and structure of DDSs (see Chapters 3–6), various data-driven storytelling processes (see Chapters 7–10), and the evaluation of DDSs (see Chapter 11).

### Chapter 2: Storytelling in the Wild: Implications for Data Storytelling

This chapter lays out the perceptual and cognitive foundation of how humans understand and perceive events, stories, data graphics, and ultimately visual storytelling. Creating meaningful and memorable stories based on data is challenging, a craft that brings together disparate strands of inquiry. There is the structure and understanding of events from which stories are crafted. There is the structure of stories and related kinds of discourse, descriptions, explanations, and conversations. There is visual storytelling, ancient and modern. There is the understanding of memory for and uses of graphic displays. Then there are the constraints of the media, print and digital, static and interactive, and the newsworthiness of the stories. This chapter points to a substantial corpus of existing research and highlights open questions and challenges for designing effective DDSs.

### Chapter 3: Exploration and Explanation in Data-Driven Storytelling

This chapter reflects on exploratory and explanatory aspects of DDSs. Exploratory aspects give readers freedom and control over how they experience the story, while explanatory aspects provide context and interpretation for the reader and allow the authors to communicate a particular narrative. The authors argue that DDSs can have high amounts of both explanation and exploration. To this end, the authors view data stories through multiple dimensions. They examine the flexibility and interpretation provided in the data stories' view, the data they focus on, and the sequences in which the data can be viewed. Examples from DDSs are used to illustrate how differing amounts of exploration and explanation can be provided in practice. Viewing these as complementary aspects could lead to new ways of integrating exploration and explanation in DDSs.

## Chapter 4: Data-Driven Storytelling Techniques: Analysis of a Curated Collection of Visual Stories

Integrating data visualization into narrative stories has now become commonplace. Authors are enabling new reader experiences, such as linking textual narrative and data visualizations through dynamic queries embedded in the text. Novel means of communicating position and navigating within the narrative also have emerged, such as utilizing scrolling to advance narration and initiate animations. This chapter advances the study of narrative visualization through an analysis of a curated collection of DDSs shared on the Web. Drawing from the results of this analysis, it presents a set of techniques being employed in these examples, organized under four high-level categories that help authors tell stories in creative ways: communicating narrative and explaining data, linking separated story elements, enhancing structure and navigation, and providing controlled exploration. The benefits of each storytelling technique along with a number of example applications of the ideas in DDSs are discussed. Furthermore, the evolution of the field and areas for future research are outlined.

## Chapter 5: Narrative Design Patterns for Data-Driven Storytelling

This chapter introduces the concept of narrative design patterns that aim to facilitate the shaping of compelling DDSs. There are many different ways storytellers can narrate the same story, depending on their intentions and their audience. Here, the authors define and describe a set of these narrative design patterns that can be used on their own or in combination to tell data stories in a myriad of ways. The authors then analyze 18 of them, and illustrate how these patterns can help storytellers think about the stories they want to tell and the best ways to narrate them. Each pattern has a specific purpose, for example, engaging the audience, evoking empathy, or creating flow and rhythm in the story. The authors assume storytellers already know what story they want to tell, why they want to tell it, and who they want to tell it to. These patterns may not only facilitate the process of creating compelling narratives, but stimulate a wider discussion on techniques and practices for data-driven storytelling.

## Chapter 6: Watches to Augmented Reality: Devices and Gadgets for Data-Driven Storytelling

This chapter discusses different device form factors and their affordances and characteristics for different storytelling settings. A wide range of form factors for data-driven storytelling, including not only the obvious electronic "devices," but also more diverse media such as tangible props (i.e., things that people can pick up and hold, gesticulate with, and so on) are considered. The latter are worth considering because they can give insights into how data storytelling might occur in futuristic mixed-reality digital environments that may be enabled by the current rapid progress in virtual and augmented reality display and interaction technologies. In addition, this discussion of devices considered not only display contents but also the possibilities for direct interaction. The chapter also presents a set of examples that use different devices in data-driven storytelling, reflecting on how to tell DDSs when using different devices and media.

## Chapter 7: From Analysis to Communication: Supporting the Lifecycle of a Story

This chapter describes the lifecycle of visual data stories, including the tools and methods that authors employ to create visual stories, the processes they go through, and the major pain points they experience. The discussion of current practices, as presented, is based on interviews with nine professional data storytellers. Each of these interviews explored the participant's experience with one past story production project. The chapter focuses on the visual data storytelling process, from conception through production, including data collection and preparation, data analysis, story development, and visual presentation. Also included is a detailed description of the main roles and activities that storytellers engage in as they turn raw data into a visually shared story and the tools they use to support their work. Based on the example projects described by the participants, the process of story production is summarized, an overview of the tools that are in use is given, and opportunities for research and design are detailed.

## Chapter 8: Organizing the Work of Data-Driven Visual Storytelling

One of the challenges of telling compelling, effective DDSs is how a group organizes their teams, skillsets, and workflow. In this chapter, we explore different approaches to working with data and visualizing data in three sectors: design studios, media, and nonprofits and nongovernmental organizations. This chapter explores what teams, tools, and organizational structures these groups use to work with data and tell narrative stories. Their experiences and lessons learned can provide valuable insight for other organizations as they seek to develop their own workflow to effectively visualize and communicate their data, analysis, and stories.

## Chapter 9: Communicating Data to an Audience

Communicating data in an effective and efficient story requires the content author to recognize the needs, goals, and knowledge of the intended audience. Do we, the authors, need to explain how a chart works? It depends on the audience. Does the data need to be traced back to its source? Depends on the audience. Can we skip obvious patterns and correlations and dive right into the deeper points? Depends on the audience. And so on. Thus, to effectively communicate ideas and concepts, content authors need to think carefully about how their work best fits the needs of the audience. This chapter explores design considerations relating to audience knowledge and goal contexts, and considers the difference between the theory of what we might know and the reality of what we can know.

## Chapter 10: Ethics in Data-Driven Visual Storytelling

Many questions that relate to ethics arise in data-driven storytelling. Is the sample representative, have we thought of the bias of whoever collected or aggregated the data, can we extract a certain conclusion from the dataset, is it implying something the data does not cover, does the visual device, or the interaction, or the animation affect the interpretation that the audience can have of the story? These are questions that anyone that has produced or edited a data-driven visual story has, or at least should have, been confronted with. After introducing the reasons and implications of ethics in this book, this chapter looks at the risks, caveats, and considerations at every step of the process, from the collection/acquisition of the data to the analysis, presentation, and publication, with many points illustrated through an example of an ethical consideration.

## Chapter 11: Evaluating Data-Driven Stories and Storytelling Tools

This chapter provides a review of how DDSs and the tools used to produce them are evaluated. Evaluation is a far-reaching concept; among the topics discussed in this chapter include the evaluation of a DDS in a newsroom context as well as the evaluation of novel storytelling tools and techniques in academic research settings. The discussion spans a diverse set of goals, acknowledging the different perspectives of storytellers, publishers, readers, tool builders, and researchers. It also reviews possible criteria for assessing whether these goals are met, as well as evaluation methods and metrics that address these criteria. This chapter is intended to serve as a guide for those considering whether and how they should evaluate the stories they produce or the storytelling tools or techniques that they develop.

## FUTURE DIRECTIONS

Many questions arise as interactive visualizations are used in situations beyond data exploration by data experts, such as the focus in this book on communication to a broader audience. Research on the understanding of static images in cognitive psychology and perception may need to be extended to encompass more advanced techniques (videos and interactive applications). Visualization literacy, or the ability to extract, interpret, and make meaning from information presented in the form of an interactive data visualization, is also a crucial component in data-driven storytelling. Assessing the visualization literacy of an audience and developing techniques to better teach how to decode interactive visualizations has started to attract the attention of our research community.* Still, there remains considerable exciting research that can to be done to contribute to a well-informed society. For example, research on how visualizations can lie (Tufte 1984) or at least how they may introduce bias in the reader's mind has focused on static visual representations. Now, opportunities, that are perhaps essential, are developing for research on the process of understanding the effects of interactivity on how interpretation emerges. Similarly, it is crucial for advancing research in visualization to assess the role data-driven storytelling can play in easing the comprehension of messages or in increasing their memorability.

Another future direction for research regards the evolution of our society in light of trends in data. Our society has entered a data-driven era in which not only are enormous amounts of data being generated every day, but also growing expectations are placed on their analysis (Thomas and Cook 2005). Analyzing these massive and complex datasets is essential to making new discoveries, communicating them, and creating benefits for people. In regards to this data deluge, what remains constant is our own cognitive ability to make sense of the data and make reliable, informed decisions. In the future, data-driven storytelling techniques will still be applicable even with growing data size and integrate advanced machine learning and data mining algorithms. New devices and interactive visualization systems will provide tools to support big-data storytelling.

---

* EuroVis 2014 Workshop: Towards Visualization Literacy. https://www.kth.se/profile/marior/page/eurovis-2014-workshop-towards-visualiza/.

# REFERENCES

Bertin, Jacques. *Semiology of Graphics: Diagrams, Networks, Maps*. Translated by William J. Berg. Madison: University of Wisconsin Press (translation from French 1967 edition), 1983.

Bostock, Michael, Vadim Ogievetsky, and Jeffrey Heer. "D3 Data-Driven Documents." *IEEE Transactions on Visualization and Computer Graphics* 17, no. 12 (December 2011): 2301–9.

Card, Stuart K., Jock D. Mackinlay, and Ben Shneiderman, eds. *Readings in Information Visualization: Using Vision to Think*. San Francisco, CA: Morgan Kaufmann Publishers Inc., 1999.

Cleveland, William C., and Marylyn E. McGill. *Dynamic Graphics for Statistics*. 1st ed. Boca Raton, FL: CRC Press, Inc., 1988.

Kovach, Bill, and Tom Rosenstiel. *The Elements of Journalism, Revised and Updated 3rd Edition: What Newspeople Should Know and the Public Should Expect*. Three Rivers Press, New York, USA, 2014.

Marshack, Alexander. *The Roots of Civilization*. Mount Kisco: Colonial Hill, 1991.

Thomas, James J., and Kristin A. Cook, eds. *Illuminating the Path: The Research and Development Agenda for Visual Analytics*. Los Alamitos, CA: IEEE CS Press, 2005.

Tufte, Edward. *The Visual Display of Quantitative Information*. 1st ed. Cheshire: Graphics Press, 1984, pp. 57–69.

Tukey, John W. *Exploratory Data Analysis*. Reading, MA: Addison-Wesley, 1977.

# CHAPTER 2

# Storytelling in the Wild

## *Implications for Data Storytelling*

## Barbara Tversky

*Columbia University*

## CONTENTS

Everyone, everywhere seems to love stories. The smallest of people can be gripped by stories, even stories they don't completely understand. What accounts for their power? So many things. An easy answer: stories are life, in all its richness. But that will not do. Stories have both more than life and less than life, and are all too frequently more engrossing than life. Stories have all the elements of life, and then some: People, place, time, emotion, intention, causality, drama, suspense, mystery, and then, lurking beneath for readers to discover and rediscover, morals, magic, and meanings of all kinds in every layer. Like life, but life condensed, sharpened, embellished, and reimagined. Stories give meaning to our lives, they interpret our lives, perhaps the reason we, singly or collectively, constantly tell and retell the stories of our lives. A recent engrossing history of humankind, *Sapiens* (Harari, 2014), concluded: it is all stories.

Crafting good stories is not easy. Small children have to learn to tell them, to learn that stories are not just *and then, and then, and then*, one thing after another. How stories are condensed and sharpened and embellished and reimagined is key to their magic. Stories—good ones—are at once both unique and universal. Stories can make the familiar strange and the strange familiar (Bruner, 1987). They draw us in through the particulars and lead us out to the general. The best of stories can reach a multitude through an individual, from the personal to the collective. Herein lies the first tension between storytelling and data. Stories are about individuals and data are about generalities.

We spend endless hours crafting and recrafting stories from our own lives (cf., Bruner, 1987; Marsh & Tversky, 2004), and endless hours watching, hearing, or reading the stories of others. We can hear so many more stories than we can experience. They expand our narrow window on the world. We can see ourselves or others in them and pull insights and life lessons from them. Stories have layers of meaning, and from understanding them we can learn how to find meaning in our lives and how to live our lives meaningfully. Stories, like our lives, can be endlessly interpreted and reinterpreted.

Data, on the other hand, seem to be the antithesis of stories. If stories have people in all their richness, data have points, rich meanings reduced to numbers. If people are ambiguous, changeable, and ineffable, data are certain, static, and clear. People have desires and conflicts and ambitions and emotions; they have successes and failures; they can be inspiring or abhorrent. Data have none of that. For some of us, the readers of this chapter, data of all kinds are exciting and meaningful, an invitation to explore and to think. Yet even we, those of us inspired by data, need the data to be presented in ways that invite exploration and thought. But for many intelligent and thoughtful people, data elicit eyes glazed over no matter how they are displayed. The more reason to weave them into good stories.

Like all good words, *story* is used by many communities with many senses. There are bedtime stories and stories in novels that are typically pure fantasy; there are the stories we tell each other about our lives that are meant to bear some semblance to truth but that are filtered through our narrative voice and point of view; there are newspaper or journalism stories that are about actual events and supposed to be free of a particular narrative voice or a specific point of view; and there are the stories of politicians often called *spin*, the stories governments and companies tell to get your trust. Just as *story* is used differently in different communities, it will be nuanced differently here, consistent with the immediate larger context. So do not expect consistency in the use of the term. No one community has ownership. One rather widely accepted sense of story will be developed below to draw distinctions with contrasting forms of discourse, notably descriptions and explanations.

## PREVIEW

Before there are stories, there are the events that give rise to them, and that is where we begin, with *event perception and comprehension.*

When people view ongoing continuous events, simplified ones that can be brought into the laboratory, they segment them hierarchically. The breakpoints between event segments are identified by large changes in the physical stream that signal changes in actions, intentions, and goals. From perceiving events to *retellings of events*, the spontaneous stories of our lives that we tell our friends. Retellings embellish by adding, omitting, and exaggerating even important details; fortunately, many of these are discounted by listeners. After that, to the *kinds of discourse*, primarily *description, explanation, storytelling* (but also *argument, journalism,* and *conversation*) and an integrative form, *narrative nonfiction*. Briefly, descriptions present the appearance and structure of people, places, things, and events organized appropriately; explanations add causality; and stories may add features like a narrative voice, protagonists and antagonists, emotion, suspense, and more. Narrative nonfiction is a form that weaves together stories, descriptions and explanations, the individual and the general, the specific and the abstract, dramatically and rhythmically. Next, *visual storytelling*, ancient and modern, noting both the expressed ideas, space, time, people, animals, and objects, and how they abstracted, depicted, and connected. From there to an analysis of how graphics work: they use space and elements in it to represent a multitude of concepts, concrete and abstract, and their interrelations. That analysis leads naturally to a discussion of empirical methods for uncovering cognitive design principles for effective graphics for various domains. Afterwards, I draw insights from *comix*, a term invented by Spiegelman (2013) to describe the inventive history of combining words, depictions, and more to tell stories. Then I take a glance at gestures and a peek at the design of the world, showing how each communicates abstract thought.

Stories can be interesting in and of themselves, but stories, especially those crafted around data, are important also for what we infer, remember, and learn from them and how they affect our actions. "We" includes many audiences who read for diverse reasons. Moreover, stories can be told using diverse media. It goes without saying that designing stories around data has to take into account the background of the audience as well as the goals, qualities, and constraints of the media.

## PERCEIVING AND UNDERSTANDING EVENTS

First comes the experiencing, then the telling. Caveat: not always. For one thing, there is the planning. And the directing. Understanding how people perceive and understand events has implications for how news stories should be crafted, what to include—goals, intentions, causes—and where to cut.

Events do not arrive neatly boxed and labeled. We do that ourselves. Events arrive as a continuously fluctuating multimodal onslaught to all our senses, James' blooming buzzing confusion (James, 1890). Some of it relevant, most not. We make sense of that ever-changing diverse stream by packaging it into discrete chunks, the categories central to our existence: faces, bodies, objects, scenes, and events, categories that are so central that language is brimming with words for them, and, save events, that the brain has specialized areas for processing them. And then to interpretations, emotions, reasons, causes, and consequences. Events take place over time, unique in that set of fundamental categories. Events stretch from quotidian ones like making a bed or doing the dishes, to epic ones like weddings, revolutions, and recessions.

How do we perceive and understand events? Several fundamental questions arise, many analogous to questions addressed in the perception and cognition of objects and scenes (cf., Tversky, Zacks, & Hard, 2008). Like objects and scenes, events have both a taxonomic structure—*kinds* of objects, scenes, and events—and a partonomic structure—*parts* of objects, scenes, and events. How do we segment events into parts? It is easier to bring the quotidian events into the laboratory to study, and many of the findings undoubtedly generalize.

In what has become a standard paradigm for studying event perception and cognition, participants view videos of familiar everyday events like making a bed or doing the dishes (e. g., Newtson, 1973; Newtson & Engquist, 1976; Newtson, Engquist, & Bois, 1977; Newtson, Hairfield, Bloomingdale, & Cutino, 1987). They are asked to press a button every time, in their judgment, one event segment ends and another begins (e.g., Hard, Recchia, & Tversky, 2011; Kurby & Zacks, 2007; Tversky & Zacks, 2013; Tversky, Zacks & Hard, 2008; Zacks, Speer, Swallow, Braver, & Reynolds, 2007; Zacks, & Tversky, 2001; Zacks, Tversky, & Iyer, 2001).

On the whole, people agree on the temporal locations of segment boundaries, referred to as *breakpoints*. When asked to segment twice, at the highest level that makes sense and at the lowest level that makes sense, people's segmentation was hierarchical, that is, the fine-level events fit nicely into the coarse-level boundaries. When asked to give play-by-play accounts of what happened in each segment as they segmented, people reported a sequence of actions on objects and completions of goals and subgoals. The higher or coarse level was segmented by different objects or object parts and the lower or fine level by different actions on the same object. Think of making a bed. The coarse-level actions were described by one participant as, "taking apart the bed; putting on the bottom sheet; putting on the other sheet; putting on the blanket…." Each new coarse-action segment involved a new object. The same participant described the fine-level actions in this way: "unfolding sheet; laying it down; putting on the top end of the sheet; putting on the bottom; straightening it out." Each new fine-action segment involved a new action on the same object.

Although these were familiar and brief events, the kind that can be completed in a few minutes and filmed from a stationary camera, a naturalistic study observing behavior in the real world found similar results. In that study, coders recorded what people in a small town were doing, day after day. Observers segmented when there was a change in person, place, object, or action (Barker, 1963; Barker & Wright, 1954). Although the events in these studies were performed by humans, not natural events like eclipses, hurricanes, or earthquakes, the latter are also perceived to have a sequence of steps associated with causes (e. g., Buehner & Cheng, 2005).

Breakpoints are privileged, marking the convergence of many cognitive and perceptual measures as well as an objective change in the sensory input, a convenient and nonaccidental concurrence. Local maxima of change in the physical stream occur at breakpoints, larger for coarse than fine breakpoints (Hard, Reccia, & Tversky, 2011; Hard, Tversky, & Lang, 2006) indicating that more is changing at breakpoints than at ordinary moments. Correspondingly, people look longer at breakpoints (Hard et al., 2011) and relevant brain activity rises at breakpoints (Speer, Zacks, & Reynolds, 2007; Zacks, Speer, & Reynolds, 2009). People understand and remember a sequence of stills composed of breakpoints better than a sequence of stills that are not breakpoints (Newtson & Engquist, 1976; Schwan & Garsoffky, 2004). Scrutiny of breakpoints reveals that they are transition points; moments where actors simultaneously complete one action and began another. Think of making a sandwich. As you are finishing spreading butter on one slice of bread, you turn your head and body to find the next slice.

Breakpoints in continuous events also mark a confluence of bottom-up and top-down processes. Bottom-up processes are moments of relatively greater change in the continuous behaviors. Top-down processes are moments where goals and subgoals are completed. That event structure is reflected in both bottom-up and top-down processes, both in perception and cognition, allows making inferences from one to the other. The ebb and flow of action enables observers of unfamiliar events to infer the hierarchical structure of events and use that to bootstrap meaning (Hard et al., 2006). Conversely, knowledge of the structure of events allows anticipating greater activity at event boundaries.

Although events are experienced as a linear sequence of actions and consequences, retellings of events can be quite different from experiencing the events, and we turn to that now.

## SPONTANEOUS RETELLINGS OF EVENTS

People talk a lot about their lives. Storytellers want to make their stories engaging, and to do so, they select what to relate and how to relate it (Dudokovic, Marsh, & Tversky, 2004), a task shared by data storytellers. To learn how that happens in everyday life, we asked university students to record events that they retold to their friends, family, and other acquaintances over the course of 4 weeks (Marsh & Tversky, 2004). For each retelling, they wrote what had actually happened, what they had said, to whom they said it, and for what reason. They wrote whether their retelling was accurate and then whether they had omitted important details, added details that did not actually happen, or exaggerated or minimized aspects of what actually occurred.

By their own admission, participants called 42% of their retellings inaccurate. But even this understated the degree to which they misreported the events they retold. Fully 61% of their retellings were altered by exaggeration, minimization, or addition or omission of important details, again by their own admission. This suggests that a considerable degree of misrepresentation is tolerated by storytellers as "accurate." An unpublished naturalistic study revealed that some degree of misrepresentation is not only tolerated but expected by listeners. If a friend laments that he did not sleep more than 4 hours a night the entire week, you might take that to mean that he is stressed out rather than taking the hours of sleep literally.

Readers probably will not be as tolerant of the distortion of events in news reports or the presentation of data. However, readers might expect such distortions in the claims of politicians or the accounts of eyewitnesses that are often significant parts of stories in the news.

## KINDS OF DISCOURSE

We begin this section with an example, perhaps the most consequential data story ever. In an email to the then graphics editor of *The New York Times*, Steve Duenes, columnist Nicholas Kristof related that Bill and Melinda Gates had changed the mission of their foundation from providing computers to third world countries to fighting disease because of columns Kristof had written in 1997 describing the severe consequences of bad water on children. What convinced the Gateses were not Kristof's words, but rather the graphic designed by Jim Perry. The graphic was simple, almost entirely text. See the mock-up below. Two columns on the left includes four diseases caused by bad water and the numbers of deaths associated with them per year. The first, diarrhea, accounted for 3,100,000 deaths per year; the other three, a total of 430,000 deaths per year. On the right column is a detailed description of the diseases. The first thing a powerful data story needs is powerful data.

Graphs, charts, and diagrams, like facts, often cannot stand on their own. They need to be embedded in some genre of discourse. Storytelling is one kind of discourse, an especially engaging form, but there are others. Even though more and more human communication is mediated, the prototypic and ancient genre is face-to-face. Face-to-face interactions are especially rich; they harmoniously blend words, prosody, gestures, and actions interactively to create meaning. Early on, people began to put thought in the world in more permanent ways, as trail markers, cave paintings, petroglyphs and

**Death by Water**

A huge range of diseases and parasites infect people because of contaminated water and food, and poor personal and domestic hygiene. Millions die, most of them children. Here are some of the deadliest water-related disorders.

| DISORDER/ESTIMATED DEATHS PER YEAR | |
| --- | --- |
| DIARRHEA<br>**3,100,000** | Diarrhea is itself not a disease but is a symptom of an underlying problem, usually the result of ingesting contaminated food or water. In children, diarrhea can cause severe, and potentially fatal, dehydration. |
| SCHISTOSOMIASIS<br>**200,000** | A parasitic disease caused by any of three species of flukes called schistosomes and acquired from bathing in infested lakes and rivers. The infestation causes blooding, ulceration, and fibrosis (scar tissue formation) in the bladder, intestinal walls, and liver. |
| TRYPANOSOMIASIS<br>**130,000** | A disease caused by protozoan (single-celled) parasites known as trypanosomes. In Africa, trypanosomes are spread by the tsetse fly and causes sleeping sickness. After infection, the parasite multiplies and spreads to the bloodstream, lymph nodes, heart, and, eventually, the brain. |
| INTESTINAL<br>HELMINTH<br>INFECTION<br>**100,000** | An infestation by any species of parasitic worms. Worms are acquired by eating contaminated meat, by contact with soil, water containing worm larvae, or from soil contaminated by infected feces. |

*Sources: The New York Times, January 9, 1997; World Health Organization; American Medical Association Encyclopedia of Medicine.*

such, preserving the visual-spatial aspects of communication carried by a joint presence in the world, but eliminating interactivities, gestures, prosody, and words. Later, written language represented word meanings. The advent of pixels allows forging the visual-spatial and the verbal back together in fluent harmony, and that forging is key to effective data storytelling.

Communication is structured at many levels, not only the familiar syntactic, semantic, and pragmatic, but also at the more encompassing level of discourse, here referred to as *genre*. Genre reflects the mode of conveyance, primarily description, explanation, or story. Many communities have analyzed genres of discourse—linguistics, literature, rhetoric, journalism, comics, drama, filmmaking, animation, narratology, law, philosophy, sociology, psychology, and more (some, but far from enough, resources for psychology and stories: Bruner, 1987; Bower, Black, & Turner, 1979; Brewer & Lichtenstein, 1982; Mandler, 1981; Rumelhart, 1975; Trabasso, Stein, & Johnson, 1981; psychology, conversation: Clark, 1996; psychology, film: Cutting, DeLong, & Nothelfer, 2010; Magliano, Miller, & Zwaan, 2001; Zacks, 2015; film: Bordwell & Thompson, 2003; Brannigan, 1992; Murch, 2001; Tan, 1996; linguistics, discourse: Brown & Yule, 1983, Gee, 2014, Ronen, 1990, Shiffrin, Tannen, & Hamilton, 2008; Van Dijk, 1993, 2001; linguistics, stories: Chafe, 1980, 1998; computer graphics: Hullman & Diakopoulos, 2011, Hullman, Drucker, Riche, Lee, Fisher, & Adar, 2013; Segal & Heer, 2010; comics, graphic narratives: Eisner, 1985; McCloud, 1994; narratology: Bruner, 1987; Prince, 2003; philosophy, Currie, 2010).

These communities have different aims and methods and produce different analyses. It is not feasible to review that work here. Instead, I present an analysis of the core genres of discourse that seem relevant here based on portions of the vast literatures.

Each genre can be expressed in words or graphics, or ideally, a combination of both, complementing each other and doing what each does best. There is an important point here. Pure words are meant to be read in order, and often are. Graphics typically do not demand or even guide an order of taking in the information. The absence of a guiding order can be confusing to some readers, doubly confusing when words and graphics are combined, and triply when interactive. How much to guide a reader and how to do that are important design considerations.

Of the core genres of discourse, three, *description*, *explanation*, and *story*, are the most central; and to these I add three more specialized but relevant forms, *argument*, *journalism*, and *conversation*. I have no doubt that experts in the various fields might take issue with some distinctions as well as also with each other. As I structure the first three forms, each adds elements to the previous form so that they build on one another. I end the section with a discussion of literary journalism, or literary nonfiction, the genre that seems most pertinent to current nonfiction including data storytelling, a form that weaves together the discourse types rhythmically.

Whether genres are pure or mixed, words or graphics, ordered or not, every piece of discourse requires authors to select the information that is relevant, to express it felicitously, and to link the pieces into a whole. An elementary decision for authors is whether to impose an order, canonical for purely verbal media, or to create interactivity, allowing readers to explore. When ordered, as in words or words and graphics embedded in a narrative, there is a beginning, middle, and end. The beginning serves as an entryway or introduction. The middle consists of a compelling sequence of related assertions. Note here the tension between surprise and drama on the one hand and continuity and clarity on the other. Jumping to a new topic without a segue might baffle readers, but also involves them, creating a mystery they are invited to solve. The end draws together and concludes a set of ideas in a meaningful way. Creativity is needed throughout. Beginnings should intrigue readers to draw them in. Beginnings may also establish expectations about the rest of the discourse, providing rubrics or drawers for organizing what follows. Endings should at the same time tie things together with a kick and leave readers with much to ponder.

## Description

A representation of the appearance and/or structure of a person, thing, space, event, or state of affairs. Each of these can use words, graphics, or both. Photos and sketches can be apt for people, places, things; so can maps, charts, and diagrams. Events can be conveyed by timelines, storyboards, step-by-step diagrams, animations/simulations, comix, and more. If in words, the order of telling is typically driven by the content described, for example, a hierarchy of importance, beginning with salient or general features and proceeding into greater detail. When describing a space or environment, people typically begin with an overview or with an entrance (e.g., Taylor & Tversky, 1992).

## Explanation

Provides an interrelated set of reasons, typically causes or justifications, for some state of affairs. The order of telling is typically driven by the temporal or causal structure and the end is typically the outcome of the process. Timelines, flowcharts, step-by-step diagrams, storyboards/comix are especially appropriate graphics for explanations. Animations must be designed with care for several reasons. People often understand and explain events over time as sequences of discrete steps, not as continuous unsegmented action. Animations all too often present too much too quickly for viewers to comprehend. Finally, animations must explain, not simply show (Tversky, Morrison, & Betrancourt, 2002).

## Stories

Typically (but not necessarily) add characters, a narrative voice, problems and resolutions, motives, intentions, emotions, suspense, and drama. Although stories typically have a temporal, indeed causal, order, they are often told in other orders to create suspense or mystery to draw readers in. Following the Russian formalists and then the French structuralists, it is common to distinguish the actual events (*fabula*) from the way the events are structured in the narrative (*syuzhet)* (cf., Bruner, 1985). Another important point—the same sequence of events, the core of a story, can be arranged in many ways; doing so serves as warm-up exercises for storytellers of all kinds (e.g., Madden, 2005; Queneau, 1947). Some arrangements will be better than others, but like wine, paintings, and lovers, which versions are best is a matter of taste.

People love to find patterns and classify, and stories are no exception. Authors of how-to guides classify story types, claiming various numbers and types of basic plots, 1, 3, 7, 20, 36, and undoubtedly more. There are the timeless plots, the stuff of myths, Greek tragedies, soap operas, scriptures, Shakespeare, *Star Wars*, and *Harry Potter*: pride before a fall; quest—for love, fame, fortune, enlightenment, power; coming of age; crime and punishment; remorse and salvation; frog into prince; hero overcomes tyrant. They have morals and life lessons: persevere and you will succeed; good triumphs over evil; love lost is better lost; appearances can be deceiving. Some of the themes are contradictory, keeping readers in suspense; if the hero gets in trouble, you do not know if his or her hubris will lead to a fall or if his or her persistence will lead to success. Stories have rhythms: the classic Aristotelean triangle of slowly rising action to crisis, followed by release and resolution. Briefer rhythms on the way: problems encountered and resolved.

Adding graphics and basing stories in data rather than individuals complicates matters of selection, expression, arranging, interactivity, mode, and genre. Recent surveys of data-telling stories have analyzed the categories that authors seem to have used. Considering formats, Segal and Heer (2010) distinguished: magazine styles, annotated charts, partitioned posters, flowcharts, comic strips, slide shows, and film/video/animations. They took note of graphic narrative devices that introduce (overview establishing shot), direct attention (e.g., grouping, arrows), and maintain orientation (e.g., visual anaphora, slider bar) as a narrative proceeds. Hullman and Diakopoulos (2011) survey a large number of illuminating examples that illustrate rhetorical devices adopted in data storytelling. Hullman et al. (2013) examined the kinds and frequency of transitions in data videos, relating them to techniques used in film. Not surprising, the most frequent transition was chronological, appropriate for descriptions and explanations; the second was general to specific, appropriate for descriptions.

Descriptions, explanations, and stories provide mental models that help us to understand the world: descriptions for the appearance and structures of organisms, things, places, and institutions; explanations for how things, people, nations, systems operate and work; stories for all forms and interactions of humanity, emotion, ambition, and desire.

Three additional forms of discourse that fit sideways rather than building on the previous are discussed below.

## Argument

An argument is similar to an explanation in providing reasons, causes, or justifications. Presumably called an argument rather than an explanation because the reasons, causes, or justifications may be open to debate and/or because arguments may provide reasons, causes, or justifications for what might happen or should be.

## Journalism

True events that are newsworthy. The critical information is: who, what, where, when, why, and how. Told first succinctly, then in more detail, and then again with related detail, represented by the emblematic inverted triangle. This triangle is a static one representing the structure of an article, from top to bottom.

Interestingly, the journalism triangle more or less corresponds to the mantra of interactive graphics: overview first, zoom and filter, and details on demand (Shneiderman, 1996).

## Conversation

Conversation is interactive. Two or more participants, each making contributions, typically brief and in alternation, that build on each other. The

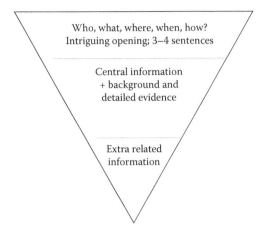

contributions can be descriptions, explanations, arguments, stories, or other. The participants do not have to be human or use words.

Although the elements of descriptions, explanations, stories, and conversations differ, in actual practice, the genres are usually mixed. Explanations typically include descriptions and stories usually include both, and arguments, journalism, and conversations can include all. This brings us to a genre known variously as literary nonfiction, literary journalism, or creative nonfiction (e. g., Buell, 1980; McPhee, 2013; Sims, 1984; Sims & Kramer, 1995; Winterowd, 1990). In today's lingo this genre would be called disruptive as it violates chronological or causal order. It borrows from literature and implicitly from music and sermons, creating the drama, suspense, variety, and change of pace to keep readers reading. It interweaves the specific and the abstract, the particular and the general, in a spiral of changing rhythm.

Implicit throughout this volume is the assumption that data journalism will be experienced on a screen or in print. As if that were not enough, virtual reality (VR), augmented reality (AR), and situated visualizations are providing exciting new ways for people to experience data stories along with new design challenges.

## GRAPHIC DESCRIPTIONS, EXPLANATIONS, AND STORYTELLING

There are many concepts that are difficult if not impossible to express visually, like *truth*, *faith*, and counterfactuals, but there many others that are expressed more directly in depictions than in language: faces, places, animals, objects, and shapes among them, as well as spaces, as in maps, events, behavior, processes, and more. We quickly run out of words that can distinguish one face from hundreds of others. Intricate actions and subtle emotions are so often better expressed by face and body than by words. Similarly, for more abstract information: line graphs showing changes in population over time, flow of water or electricity, chemical bonding, or mechanics of, for example, car brakes. As the billions of consumers of graphic novels, film, and video know, visual stories can be especially powerful, captivating, and memorable. More and more, media combine both words and images. But storytelling began as depictions long before there were written languages.

## Ancient Graphics

Graphic representations are ancient, long preceding written language. Frequent topics were beings and things, space, time, and number. Vivid images of animals, singly, in groups, and in stampedes, remarkably still adorn the walls of the Chauvet caves in the south of France, even though they were painted at least 35,000 years ago. Animals, along with handprints and more rarely, people, appear on walls of ancient caves and surfaces of petroglyphs all over the world. Ancient maps, those that survived, typically mixed perspectives with overviews of paths and rivers and frontal views of landmarks, buildings, mountains, and the like. Maps of heavenly bodies appear in ancient times and even earlier, and situated visualizations like sundials and Stonehenge and Mayan temples that were carefully aligned with paths of celestial bodies. Processions of images painted on the walls of Egyptian tombs more than 3000 years ago show events in time step-by-step: how to grow and harvest wheat and make bread or how to make bricks (Smith & Simpson, 1998). Time was visualized more abstractly in calendars, often circular. Ancient tallies inscribed in bones or stone have been discovered in remote parts of the world dating back more than 35,000 years. Some seemed to have been used for accounting, others for astrological recording (Pickover, 2009). In fact, writing is regarded to have begun for accounting, that is, for keeping track of data—the sizes of herds of cattle and sheep for taxation (Schmandt-Besserat, 1992). Ancient mandalas and friezes scattered across the eastern world show the panoply of Buddhist and Hindu gods. Aztec and Maya codices depicted their unfolding history, often superimposed on a map (Hassig, 2001; Sharer & Traxler, 2006). Greek and Etruscan vases, mosaics, friezes, and frescoes illustrate their myths and histories. As for many ancient depictions, relative sizes and positions often do not reflect actual size or position; instead, they have symbolic meanings, reflecting power or social rankings (Small, 2015).

These marvelous and varied remnants of ancient cultures appear to intend to show, describe, record, explain, and inspire. They depict concrete things, people, animals, and tools; they portray things in space and show events in time; they represent numbers for various uses; and some seem to be purely spiritual.

## Modern Graphics

Intriguingly, graphics that do more than show or summarize observable entities and events are nearly absent before the late 18th century. William Playfair in Scotland and Johann Heinrich Lambert in Switzerland (Beninger & Robyn, 1978; Spence, 2000) are credited for developing the first displays of data, line graphs showing changes in some value, usually economic, over time. It is probably the most common data visualization to this day. At that time, diverse scientists and engineers invented new ways of abstracting and depicting data and processes (see http://www.datavis.ca/milestones/index.php?group=1700s, a website developed and maintained by Michael Friendly and Daniel J. Denis). These visualizations became tools of thought and inference for scientists and policy makers alike. Their appearance not only coincided with advances in science and technology, but spurred those advances. The science and technology and the visualization of science, technology, and data leapfrogged each other, a virtuous cycle that continues—think, for example, of Feynman diagrams, cloud chambers, the Hubble telescope, the tunneling microscope, and the double helix. Modern graphics include displays of data and diagrams of mechanical, astronomical, anatomical, and other scientific processes, many of which are not directly observable.

Unlike ancient graphics, modern ones use words and symbols as well as visual-spatial representations of information. The use of words in visualizations is quite different from their use in text. In graphics, the visuals are primary, the words annotated. Words do not appear in sentences; they are used to label, augment, and clarify the visual information. Similarly, numbers and symbols are incorporated into graphics.

Information diagrams began to appear in abundance in the same era, perhaps promoted by the publication and popularity of Diderot and D'Alembert's *Encyclopedie* (1751/1772) (see Figure 2.1). Purporting to be a compendium of all accumulated human knowledge in the sciences, arts,

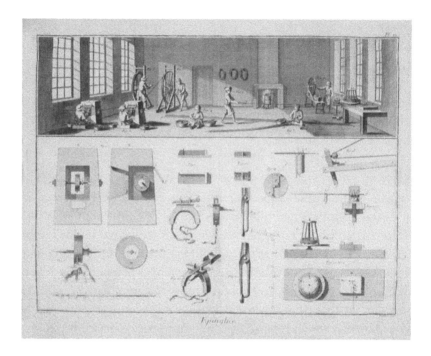

FIGURE 2.1 The pin factory plate from the Encyclopedie of Diderot and D'Alembert.

*Source:* **https://commons.wikimedia.org/wiki/File:1762_Diderot%27s_ Encyclopedie,_Epinglier_II.jpg.**

and industry, the encyclopedia opened with a conceptual diagram, a tree of knowledge divided into memory, reason, and imagination. The 28 volumes with over 70,000 articles written by diverse contributors contain over 3000 diagrams. Many of the diagrams have the same format, a format that teaches readers what a diagram is. There are two halves, each enclosed in a box. The boxes are containers and separators: they contain one kind of information display and separate that kind from the other kind. The top halves depict a natural scene, typically an industry workshop of some sort—dressmaker, beekeeper, glassmaker—with workers using the instruments of their trade. The scene is caught in action; the people holding the instruments they use proportional to their actual sizes and in the appropriate places and positions, and the light and shadows determined by the light coming from the windows. Scenes would have been familiar to readers at the time, but the format of the information display in the bottom half would have been novel. The bottom halves show only the tools and instruments of the trade. They are arranged like a catalog in rows and columns, grouped by related functions. Their image sizes are more or less equal, not proportional to their actual sizes. Their orientations and shading are devised to enhance their 3-dimensional (3-D) structure, not to reflect ambient lighting. Both parts are enclosed in a larger box, separating the diagram from the surrounding text. It is as if the diagram is telling you that it is a diagram as well as teaching you what a diagram is. A diagram was a new form of representing information, in contrast to a familiar form, a depicted scene. Diagrams do more than depict; as Bender and Marrinan (2010) note, they are meant to be "worked with"; that is, studied for inference and understanding.

Other influential visualizations in the 19th and early 20th centuries that were meant to be worked with include Snow's map of cholera cases in London (see Figure 2.2), Minard's visualization of Napoleon's unsuccessful campaign against Russia (see Figure 2.3), Nightingale's rose (circular histogram) diagram of the causes of mortality in the Crimean War

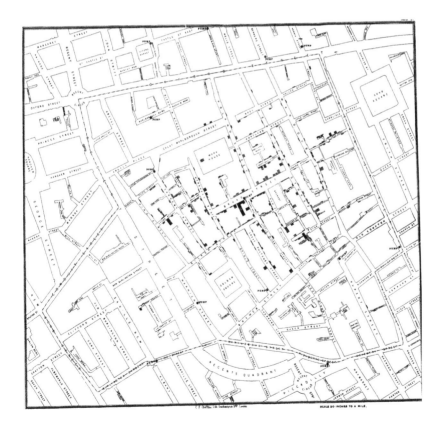

FIGURE 2.2   Snow's map of cholera epidemic.

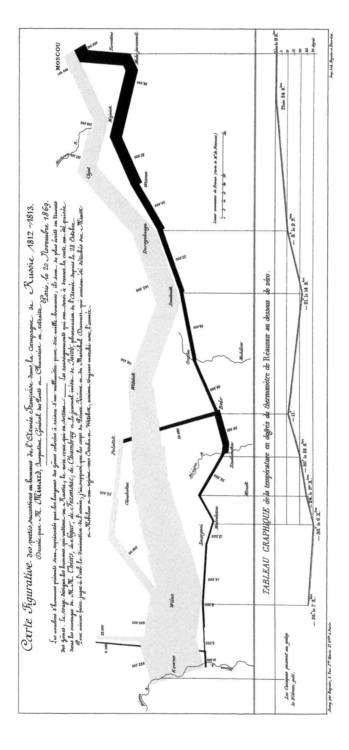

FIGURE 2.3   Minard's diagram of Napoleon's campaign against Russia.

*Source:*   https://en.wikipedia.org/wiki/File:Minard.png.

(see Figure 2.4). Snow's map showing a cluster of cases of cholera at the Broad St. pump in London supported the inference to remove the handle of the pump; this led to the abatement of the epidemic, even prior to the germ theory of disease. Displaying cases on a map is in wide use today in analyzing epidemics, migration patterns, voting patterns, and more (see Figure 2.2). Minard's visualization of Napoleon's campaign used diagrammatic space with several different meanings: to represent geographic space, to represent movement in geographic space, to represent the size of troops, and to represent temperature (see Figure 2.3). Artful design prevents confusion in interpretation. Nightingale's visualization vividly showed the surprising finding that more far more military deaths were due to disease than to battle (see Figure 2.4).

The Vienna Circle, a group of philosophers, logicians, and scientists active in pre-WWII Vienna, was devoted to universals of science and accessibility of ideas; part of that concern was expressed in developing Esperanto as well as pictorial expressions of meaning, including picture languages and isotypes for data. Isotypes were developed by Otto Neurath (1936). An icon, for example, a sack of wheat or a factory represented a specific quantity of the corresponding item. The icons were arranged in rows or columns like bar graphs (see Figure 2.5).

## Contemporary Graphics

The explosion of creative contemporary graphics has been exhilarating; they are everywhere. The digital age has enabled the explosion; everything is pixels, words and images, verbal representation and visual-spatial representation are conjoined, just as words and gestures are conjoined in face-to-face communication. Especially important for the current discussion are developments in graphic instructions, explanations, infographics, and comics/comix/graphic novels which now extend beyond fiction to nonfiction, science, journalism, Ph.D. dissertations, and much more. Inspiring innovations will continue to appear, making it impossible to collect a comprehensive list. At the same time, VR and AR are already producing inventive ways to visualize information and promising even more exciting developments (much is on websites; here are a few of the many print resources: Bertin, 1981; Card, Mackinlay & Shneiderman, 1999; Cleveland, 1985, 1993; Few, 2012, 2013; Fry, 2007; Heer & Shneiderman, 2012; Kosslyn, 2006; Larkin & Simon, 1987; McCloud, 1994; Munzner 2014; Tufte, 1983, 1990, 1997; Viegas & Wattenberg, 2007; Ware, 2013).

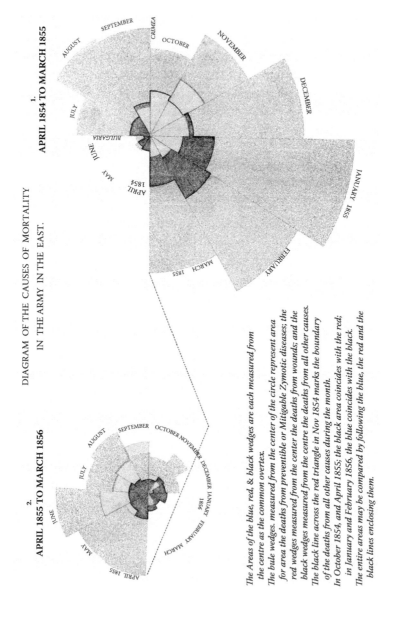

DIAGRAM OF THE CAUSES OF MORTALITY
IN THE ARMY IN THE EAST.

FIGURE 2.4   Nightingale's rose of Crimean War casualities.

*Source:*   **https://commons.wikimedia.org/wiki/File:Nightingale-mortality.jpg.**

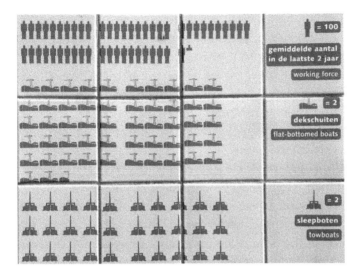

FIGURE 2.5   Neurath Isotypes.

*Source:*   **https://commons.wikimedia.org/wiki/File:Isotype-neurath.jpg.**

## HOW GRAPHICS WORK

Graphics of all kinds—maps, diagrams, charts, graphs, and sketches—use marks and places in space to convey meanings more directly than words. These meanings are carried in gestures, the precursors of graphics. In fact, much can be learned for designing effective graphics from eliciting gestures. The analysis presented here is taken from previous works (expanded in Tversky, 2011a, b; 2015; Tversky & Kessell, 2014). It might seem audacious to claim that the uses of space and marks are used and interpreted meaningfully, but each claim is supported by data, some historic, some from current best practices, or some from empirical research.

### Space

There are three core ways that places in space are used meaningfully—proximity, centrality, and direction. These uses can be regarded as design principles. First, proximity in space represents proximity on any dimension. Proximity is, of course, a fundamental Gestalt principle of spatial organization (e.g., Hochberg, 2014). Things more related on any dimension are closer in space; conversely, things that are closer spatially will be interpreted as more related conceptually. This is most emphatically exemplified in maps. Proximity is not only natural in many cases, like maps, but is also a design principle for any visualization; and, for that matter, any communication: put spatially together information that is conceptually close (e.g., Larkin & Simon, 1987). A corollary of proximity is grouping: things that are grouped together are more similar to each other than things in other groups. Grouping provides a basis for categories and hierarchies. Members of well-defined groups are treated as equivalent on some dimension or attribute.

The next design principle is *place* where the center and periphery represent their metaphoric selves.

The third design principle is *direction*. The vertical dimension is powerful, and indeed represents power. Going up entails overcoming gravity; that requires resources: age, strength, health, power, and money. Hence, all good (or desirable) things tend to go up. This is reflected in language—she is at the top of the heap or he falls into a depression. It is also reflected in gestures like *thumbs up* (or *down*) or *high five*. Many expressions and gestures differ across cultures, but because gravity is universal, the association of *up* with everything good holds around the globe. Except for economics, where unemployment and inflation are plotted upwards. There are exceptions: this is not to say that economists are perverse, but when the numbers conflict with the concepts and the numbers win, they increase upwards. The horizontal is more neutral; the world has a powerful vertical asymmetry but lacks horizontal asymmetry. However, bodies have some asymmetries. Among them, handedness and the dominant hand is associated with greater value (Casasanto, 2009). The vast majority of people are right-handed and this mapping is limited to value, and perhaps to value regarded categorically. A horizontal asymmetry with more pervasive effects comes from a cultural convention, a reading/writing order. Spontaneously generated increases in almost any variable, including value, correspond to a reading/writing order, especially time, a central aspect of stories, explanations, and visualizations (Schubert & Maas, 2011; Tversky, 2011a; Tversky, Kugelmass, & Winter, 1991). Combining a place in space and reading order yields two more principles. Indentation, as seen in outlines, indicating subcategories or subimportance, as does lower, as in trees or hierarchies with roots at the top.

### Marks

Marks come in many varieties. They can be depictions, photos or drawings of politicians or actors, picnic tables on road signs, waste baskets on computer docks, rows of factories in isotype bar graphs, etc. Marks also include depictions of churches or campsites on maps; they include depictions that bear metaphoric relations to what they represent, a crown for a ruler, Big Ben for London, the Eiffel Tower for Paris, etc. Written languages started as depictive. Marks can be words and symbols as well, but depictions have an enormous advantage over words in accessibility to meaning, in distinctiveness, and in memory. The Gestalt principles enter here as well: similar marks indicate similar concepts; larger, more salient marks indicate more important concepts.

## Meaningful Schematic Marks

More intriguing are the meanings of marks like dots, lines, blobs, and arrows. These schematic marks carry meanings that depend on their Gestalt or geometric properties and that are typically clear in contexts (and have parallels in gestures). Think of route sketch maps. Lines typically represent paths, dots typically represent places. More generally, dots represent ideas and lines the relations between ideas. Lines and dots form the basis of networks, where the dots may represent computers or friends or ideas depending on the kind of network, and the lines represent the relations between the computers, friends, or ideas. Arrows are asymmetric lines and represent asymmetric relations whether indicating a one-way street or a one-way transformation. Blobs typically represent regions, parks, or neighborhoods on maps; processing centers in representations of brains or computer systems. As regions, they can be containers of other things—the citizens of a country on a map or sets in Venn diagrams. Dots, lines, blobs, and arrows have context-dependent meanings and form a basic visual vocabulary for creating visualizations (Tversky, 2011) as well as art (Kandinsky, 1926/2013; Klee, 1953) and language (Talmy, 1983, 1988).

## Inferences from Visualizations

Visualizations are meant to enable inferences and insights. Importantly, the ways that information is represented leads to different kinds of inferences. For example, lines draw the eyes. They connect disparate entities; they suggest that the instances on the line share an underlying dimension but differ in values on that dimension. Bars contain and separate; they suggest that things within a bar are similar or related and different from the things in other bars. In several experiments, the same data were presented either as bars or as line graphs. Those who saw bars interpreted the data as a discrete comparison and those who saw lines interpreted the data as trends (Zacks & Tversky, 1999).

Arrows also carry strong meanings that affect inferences from visualizations. Undergrads were asked to interpret diagrams of a bicycle pump, a car brake, or a pulley system (Heiser & Tversky, 2006). Half the diagrams had arrows, the other half did not. Diagrams without arrows were interpreted as structural, locating parts relative to one another. By contrast, diagrams with arrows were interpreted as functional, presenting the sequence of actions and outcomes from beginning to end. The arrows changed the meanings of the diagrams enormously. Similarly, when asked to produce diagrams from descriptions, participants given functional descriptions drew arrows, but participants given structural descriptions did not.

## DESIGNING EFFECTIVE GRAPHIC DISPLAYS

### Two General Design Principles

We have come quite naturally to issues of design of effective graphic displays of information. Trade-offs are inevitable but also informative and helpful. For clear communication to ordinary people, reduce the data to essentials; for data exploration by experts, provide more data in many forms. That is, provide the information needed for the task at hand. That is, of course, easier to say than to do as different tasks and different users have different needs.

More generally, good graphics should conform to two principles, one about the content and form of the representation and the other about the perception and comprehension by the user (Tversky et al., 2002; see also, Norman, 2013). *Congruence*: the structure and content of the external representation and should correspond to the desired structure and content of the internal representation; *Apprehension*: the structure and content of the external representation that should be readily and accurately perceived and comprehended.

A case in point is animation. Animated graphics do not stay still; they keep changing. According to Betrancourt and Tversky (2000, p. 5), "computer animation refers to any application which generates a series of frames, so that each frame appears as an alteration of the previous one, and where the sequence of frames is determined either by the designer or the user." Animations can be attractive and pleasurable in and of themselves as attested by billions of viewers of films, videos, and more. In interface design, they are used for that and for many other ends: to attract attention, as our eyes quickly attend to moving things; to keep us oriented as we move in a virtual world or as the world we are viewing moves; to show patterns of movement like the flow of fluids or schools of fish; or to present or explain behavior or processes. An especially informative survey of animation in interfaces (Chevalier, Riche, Plaisant, Chalbi, & Hurter, 2016) reduces 23 uses of animation to 5: keeping context, teaching, improving user experience, data encoding, and visual discourse.

The use of animation to teach, to explain behavior or processes, is of special interest here. Because these animations use change over time to convey change over time, they conform to the Congruence Principle (Tversky et al., 2002). However, a research survey on explanatory animations used in a multitude of contexts revealed that animated graphics did not surpass static ones when the content was equivalent and when interactivity was not allowed (Tversky et al., 2002). It is known that interactivity can have benefits but can also hinder or not be used at all, depending on the kind, content, purpose, and mode of interaction.

What accounts for the failure of so many explanatory animations to surpass their static equivalents? On the one hand, animations often show too much too fast, violating the *Apprehension Principle* (Tversky et al., 2002). But there are other disadvantages of animations. Showing is not explaining, and typically, animations only show. Animations usually show processes in proportional time, but the events underlying animations are naturally segmented into steps by changes in actions and objects, steps that rarely occur at equal time segments (Zacks et al., 2001). Moreover, processes and events that occur over time are usually explained as a sequence of discrete steps. Designing effective animations is complicated and it appears that there has been progress. A more recent meta-analysis (Bierney & Betrancourt, 2016) provides some encouragement for the use of animations in educational settings.

Well-designed animations for other purposes can be effective (Chevalier et al., 2016), for example, animations that are designed to keep viewers interested or animations intended to show patterns of motion, from fluids to fish to football. Especially effective are animations aimed at preserving orientation in space and time as we explore environments or as we observe and track changes in displays of people, things, or data, for example, from bar graphs to correlations (Heer & Robertson, 2007; Heer & Shneiderman, 2012).

## How to Find Cognitive Design Principles: The Three P's

Congruence and apprehension are good principles but do not provide guidelines for specific applications. Good designs often arise from trial and error in communities of users interacting on various tasks. Language is an example; language emerges in large and small communities and changes as needs and goals change (e.g., Clark, 1996; Donald, 1991). So too for commonly used visualizations, like sketch maps. This natural process of iterative trial and error can be shortcut in the laboratory to provide empirical methods to find and test design principles through a program we call the Three P's: Production, Preference, and Performance (Kessell & Tversky, 2011; Tversky et al., 2007).

The application of the Three P's to the design of data visualizations has provided unexpected insights as well as cognitive design principles. In one set of studies, people were asked to produce representations of people in places at different times. People primarily produced tables or line graphs that connected people to places over time. The forms of the representation elicited qualitative and quantitative differences in inferences. There were more inferences and more varied inferences from the tables than from the line graphs. For example, there were more inferences about relations among people from tables. By contrast, the line graphs encouraged inferences about movements of people (or things) across places and time. These are quite frequently the kinds of inferences desired when tracking diseases or drug trafficking (Kessell & Tversky, 2011; Tversky, Gao, Corter, Tanaka, & Nickerson, 2016). The lesson is to choose the form of representation that encourages the desirable inferences—tables for more varied inferences, lines for inferences about movement in time. In another application, people produced, preferred, and performed better when continuous relations were represented by continuous graphics and discrete relations by discrete graphics (Tversky, Corter, Yu, Mason, & Nickerson, 2012). That insight led to the successful design of a touch interface for math: discrete one-to-one gestures for addition and continuous gestures for estimation (Segal, Tversky, & Black, 2014). The general cognitive principle is one of congruence: the form of the representation or the interaction should be congruent with the form of thought.

The Three P's has been successful for the design of info graphics starting from assembly instructions; think of putting together knock-down furniture, as well as the design of data visualizations. For assembly, first one group of participants *produced* instructions after completing assembly. Another group rated *preferences* for those instructions. From this we extracted potential design principles and then tested those for efficacy of *performance* in a third group (Agrawala, Phan, Heiser, Haymaker, Klingner, Hanrahan, & Tversky, 2003; Tversky, Agrawala, Heiser, Lee,

Hanrahan, Phan, Stolte, & Daniel, 2007). The cognitive design principles that emerged have broader implications for explanations and stories, including data stories: show step-by-step, show each causal action, show perspective of action, that is, show each step in the process and show how the steps are connected.

## LOOKING FORWARD: INSIGHTS FROM COMIX

Like traditional visual explanations and stories, comix present step-by-step boxes so that these are sometimes referred to as comic book explanations. Of course, comix do so much more, as shall be seen. They differ from animations in a critical way: each new frame may not be an alteration of the previous one. In many cases, people do not need the perception of continuous motion. Filmmakers, novelists, and nonfiction authors know this well, and often jump spatially to different perspectives or places, or jump temporally to different times, even prior ones, or both. Their audiences can typically follow these cuts and eventually make sense of them.

Step-by-step presentations predate comics, going back to bread-making in Egyptian tombs, Trajan's column depicting Roman military victories, and the Bayeux Tapestry depicting the events leading up to the Norman conquest of England culminating in the Battle of Hastings. Designing step-by-step presentations requires: appropriate segmentation, appropriate presentation of each segment, and appropriate connections of the segments. Comics and graphic novels have gone far beyond step-by-step presentations. They have developed a stunning range of visual-spatial tropes for telling stories that verge on the poetic, but more accessible than poetry in words. Most of the tropes have no parallels in language and lack Greek names. These examples should provide inspiration for designers of stories relying on data or any kind of visual-spatial story or explanation. There are far more than we can present here. Sadly, indeed ironically, we are limited to descriptions here. Some comix devices have already been adopted in data storytelling (e.g., Bach, Kerracher, Wim, Hall, Carpendale, Kennedy, & Riche, 2016).

Every designer will want to keep McCloud's *Understanding Comics* (1994) close at hand. It is a comic that analyzes comics, and in doing so, provides insights into storytelling and story-understanding in general, and in the conjoint use of depictions and language to do so. Also to be kept close at hand are Spiegelman's *Co-Mix* (2013), a grab-bag of insights from one of the masters of the medium and Eisner's *Comics and Sequential Art*

(2008), more insights from another master of the medium. Studying the oeuvres of both Spiegelman and Eisner and many many other practitioners of the medium will also be rewarding. Larry Gonick's cartoon guides to many disciplines, such as physics, chemistry, genetics, calculus, history, and more, provide a cornucopia of creative and insightful visual storytelling (http://www.larrygonick.com/).

There has been lively discussions of what to call this medium and what it includes. Here we will use Spiegelman's *Co-Mix* (2013) to encompass combinations of words and graphics in comics, graphic novels, graphic journalism, graphic nonfiction, and cartoon textbooks.

### Boxes/Frames

Comix have rhythms and are typically segmented by pages, boxes, speech balloons, and speech bars. They are read from top-to-bottom and left-to-right in European languages, but right-to-left in Japanese. Groensteen's fascinating book, *The System of Comics* (2009), analyzes the structures and meanings of the hierarchical and spatial arrangements of boxes, balloons, bars, and the visual means of enclosing and categorizing some information and separating it from other information. Many artists begin with pages devoid of images with $3 \times 3$ or $3 \times 4$ arrays of boxes, and they fill those in with the story. Others prefer the freedom of using larger and smaller boxes, or no boxes at all. Some introduce a story with a splash page, a two-page spread much like an overture or a trailer, that includes many of the characters and scenes and themes that will follow. Other splash pages simply show the overall setting for the story. A frequent successful device is to overlay that kind of splash page with the typical rows of boxes that show the action of story. This allows readers to get an overview and detail at the same time.

We know the meanings of boxes: they bind and bound. They contain a set of related things and separate that set from other sets of things. Bigger boxes contain more stuff and smaller ones less. In art, rules and restrictions are meant to be artfully broken. The same applies for boxes. In a fanciful comix retelling of *The Three Pigs* (Wiesner, 2001), the first pig breaks the fourth wall. Then he walks into the gutter—the space between the frames—and peeks back into the preceding frame to tell the second pig, "Come on—it's safe out here." This trope acknowledges that the story is in the boxes, and at the same time, entices the reader into the meta-story outside the boxes. In fact, the pigs all come out and proceed to stamp on the wicked story in the boxes to the delight of readers. In another case, a character in one frame, the young physicist Richard Feynman entertaining several women at a party at Columbia, reaches out of his frame back to a character in a previous frame, his girlfriend dancing with a gentleman at MIT, to hand her a letter (Ottaviani, 2011).* Breaking the frame has also been used to defy time in science fiction: imagine a character in a later frame handing a weapon to a character in need in an earlier frame.

Frames have also been used figuratively with double meanings to great effect. In a wordless panel by the early 20th century comics artist Winsor McKay, a sequence of frames shows a child's evolving sneeze; when the sneeze bursts, the frame shatters, to the bemusement of the child. In order to show the extreme speed of a chase, Jim Harriman, the creator of the manic Krazy Kat, depicts the chase in a diagonal sequence of narrow tilted frames sweeping down the page.

## Segmenting

Unlike films, comix are highly discontinuous and abbreviated, packed into separate frames or boxes. Boxes contain and separate. They segment time and space. How should that segmentation be accomplished? And what should be depicted in each frame? The research on event segmentation and on the Three P's suggests two cognitive design principles: (1) segment by goals and subgoals and (2) show the most informative parts of the events in each segment, the breakpoints. Indeed, for visual explanations like furniture assembly, this technique seems to work well, and should work well for data storytelling. Many data stories do concern changing events over time and the designers' goals are often to tell the stories in a straightforward way.

By contrast, designers of comix stories often want drama and surprise, which demand not telling stories in a straightforward way. Time can be drawn out by showing small changes in successive frames or chopped up into jumpy, staccato changes. McCloud (1994) discusses some of visual devices that can alter the experience of time.

## Connecting: Visual Anaphora

Because successive frames in comix are typically not continuous, comix need to provide ways for readers to bridge the gutter, to understand what is happening from frame-to-frame. Language does that with anaphora.

---

* I am indebted to Jon Bresman for this and many other examples as well as hours of enlightening discussion on the art and design of comix.

Consider: The boy fled the scene recklessly. He tripped and fell. The "he" refers back to the boy, so the reader knows it's the same person. If the second sentence began with "She," the reader would look back farther to find mention of a girl—or be confused. Similarly, comix often use visual anaphora; some thing or things in frame $n$ carried over to frame $n + 1$ to let the readers understand that we are still on the same sequence of events. If many things change, we have jumped to another place or another time.

## Metaphors

Visual metaphors are common in comix, and even more so in cartoons, historically as well as contemporary. Much can be learned from studying them. In a piece for the *Los Angeles Times*, Winsor McCay depicted a businessman walking rapidly on the streets of New York; the street rolls up into a treadmill, and the man keeps up his pace. In his comic strip *Little Nemo*, McCay depicts Little Nemo asleep in bed. Soon he is transported to dreamland. To show that, McCay has the bed's legs grow longer and longer and take Nemo outside into another world. When the dream ends abruptly, as dreams do, the long-legged bed dumps Nemo back into his real one (see Figure 2.6). Visual metaphors provide rich meanings that cannot easily be conveyed in words. They are distinctive and memorable.

## Visual Jokes

Humor is always welcome: all the more when it carries deeper meaning. The metaphors described above carry humor: the street rolling up into a rat race, the bed grows long legs that transports Nemo to dreamland. To create the feeling of a mad chase, Krazy Kat racing after a huge rolling stone that is wreaking havoc, Herriman draws the frames diagonally across the page. When McCay's Little Sammy sneezes, the frame shatters (see Figure 2.7). The grandmaster of visual jokes is Saul Steinberg, whose cartoons were far more than jokes. His telescoped maps, oversized close-up and shrunken in the distance visualized our conceptual myopia, what is close looms large, what is in the distance, spatially, temporally, socially, and politically is less consequential. Steinberg shows pompous people in conversation, but the speech bubbles contain a gibberish of symbols: noise without meaning.

## Perspective

The view presented to readers can be from varying distances and varying perspectives depending on the intended meaning. Close-ups naturally

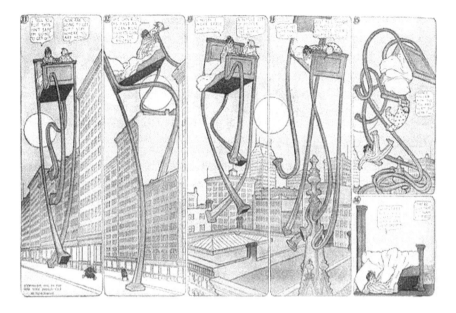

FIGURE 2.6    Little Nemo's dream by Winsor McCay.

FIGURE 2.7    Little Sammy by Winsor McCay.

give intimacy; at the same time, they ignore context. Graphic presentations can easily switch perspectives within a scene to highlight the perspective of one person or point of view or another. Similarly, graphic presentations can provide overview perspectives that create distance and give broad overviews and more equal weight to everything rather than biasing one point of view over another. Surprising perspectives, for example, from the bottom of a scene, can jar readers and make them see things anew. In visualizing demographic data, turning the world map upside down is dramatic, even confusing, but does lead to recalibration, reinterpretation, and reconsideration by audiences.

## Words, Symbols, and Pictures

Picture and words can be complementary, contradictory, redundant, or supplementary. Words and symbols themselves can be depictions. They can look loud or important by being large or thick; they can be sounds or tastes or smells or motion or vibrations. Comix can convey far more sensations than the visual and the aural. Depictions can contrast ironically or poignantly with words. A small child is listening to an adult describing a brutal murder. The adult says X was "rubbed out," a Britishism for murder. The child imagines a picture being erased, as "rubbed out" also means erase in British English (Gaiman & McKean, 1987).

### Simultaneous Parallel Stories

Comix can tell two stories simultaneously by breaking up a page into two sets of frames that are easily distinguished, say by color or theme or foreground/background. One powerful page simultaneously can show a murder in depictions and the terrifying sounds rendered into letters. Another can show the present actions of a man in the foreground and his actions in the past in the background.

### Double/Triple Meanings

We have seen that boxes and frames both contain and separate, a convenient and elegant double meaning. Our first pig stepped out of the comix world into another one. An example of a triple meaning is created by an old fashioned phone cord. A green female superhero is talking on the phone to three co-conspirators. The phone cord, representing itself, is wrapped around each of the three frames of the co-conspirators. The phone cord thereby represents the frames as well as defining, drawing together, the group of conspirators.

### Words

Comix introduce words in several different ways, using visual-spatial means to differentiate them. Speech appears in balloons with tails pointing to the speaker; thoughts also appear in balloons with the thinker indicated by a column of circles in decreasing size. Commentary or narration is typically presented as text in italics either at the top or the bottom of a frame, separated from the frame by a boundary line (see Groensteen, 2009). Talk can be expressed in multiple ways—sentences or isolated words in odd fonts and sizes, grunts, gasps, and other noises. Commentary appears in canonical sentences. These conventions help readers interpret the many levels operative in comix, but like all the conventions, they can be broken for effect.

### A Caveat on Culture and Language

Both culture and language appear to affect qualities beyond content of depictions in comix, picture books, and information visualization. The street scenes in Eastern cultures are judged more complex than those in the United States (e.g., Miyamoto, Nisbett, & Masuda, 2006). Correspondingly, covers of popular Korean picture books were judged to be more complex and convey more action than covers of popular U.S. picture books (Won, 2011), and eastern comics were judged to portray more action than western (Tversky & Chou, 2010). Similarly, popular Chinese visualizations of *addition*, *subtraction*, *multiplication* were rated as more complex than U.S. visualizations (Wang, 2011; Zhang, 2014). Designers may need to take these differences into account (Tversky & Chow, 2017).

## RETURNING TO DATA

We have not forgotten those stripped-down points anchored in various information spaces. Fundamental questions for a data journalist crafting a story are: what data are relevant? How should they be displayed? Data can be quantitative, those stripped-down points, or qualitative, beginning with eyewitness accounts (n.b., these are often not reliable). Spatial/structural data lend themselves to descriptions, data such as patterns of disease, crime, housing prices, or organizations of governments and corporations. Temporal/causal data lend themselves to explanations and stories, the actions of individuals, changes in gross national product (GNP), population, or education over time. Spatial/structural data invite maps and organizational charts; temporal/causal data invite flowcharts and line graphs. Many creative designers have developed beautiful graphics that are tempting to use, but if everyone adopts the same displays, they will be confused in memory. Yet another trade-off. One goal of data storytellers is to make their stories distinctive, individual, and memorable. Another goal is to insure that the displays encourage the desired (and true) inferences and conclusions intended by designers.

## DESIGN OF THE WORLD: *SPRACTION*

From the design of graphics to the design of the world, to *spraction* (Tversky, 2011). So little of the world is wild. Our actions design the world, the space around us: into rows and columns on shelves and buildings and streets that reflect categories and hierarchies and orders; into rooms and buildings with themes; into table settings and buildings with 1-1 correspondences and repetitions and symmetries. The actions that create the organizations, putting, ordering, lifting, turning, separating, and more, get abbreviated into gestures that communicate those actions, literally or metaphorically. Those words are used to describe our thinking, our mental actions on thoughts. The patterns created by the conceptual organizations form templates for graphs and charts: bar charts, matrices, and line graphs. These ways of organizing the world represent highly abstract concepts. The patterns created are good Gestalts that attract the eye and invite reflection: what do they mean? Spraction: actions in space express abstractions.

## PULLING THINGS TOGETHER

We began with the observation that stories have universal appeal, that they are like life, but better. We have taken a tour of the many layers that contribute to storytelling, to visual-storytelling, and to visual data storytelling including understanding events, retelling events, creating visual-spatial meanings, and discovering cognitive design principles for effective graphics. Communication first developed face-to-face, and communication in the wild uses many layers and modes simultaneously—words, prosody, gestures, actions, and props in the world. The freezing of communication into depictions in caves, on stone, in stained glass windows, and words on tablets and paper allowed preserving and disseminating knowledge, but it also led to separating the natural layers and modes and even eliminating some of them, notably prosody and gesture. Contemporary visual explanations, data journalism, and comix reunite depictions, graphics, and language and even prosody and gesture in rich and varied visual forms; they expand the form and enlarge our understanding with a wealth of delightful visual tropes.

## ACKNOWLEDGMENTS

I am deeply grateful to Christoph Hurter, Nathalie Riche, and Pat Hanrahan for excellent suggestions and guidance on many aspects of the chapter, to Marguerite Holloway and Scott Slovic for discussions of narrative nonfiction/literary journalism, and to my collaborators on many of the projects described. I am also grateful to the following grants that provided direct or indirect support to the research described or to the preparation of the manuscript: NSF CHS-1513841, NSF HHC 0905417, NSF IIS-0725223, NSF IIS-0855995, NSF REC 0440103, the Stanford Regional Visualization and Analysis Center, ONR NOOO14-PP-1-O649, and The John Templeton Foundation through The Varieties of Understanding Project at Fordham University. The opinions expressed here are those of the author and do not necessarily reflect the views of The Varieties of Understanding Project, Fordham University, or The John Templeton Foundation.

## REFERENCES

Agrawala, M., Phan, D., Heiser, J., Haymaker, J., Klingner, J., Hanrahan, P., & Tversky, B. (2003). Designing effective step-by-step assembly instructions. In *Proceedings of SIGGRAPH 2003.* ACM Transactions on Graphics, pp. 929–937.

Al-Biruni, R. (1029/1934/2006) *The book of instructions in the elements of the art of astrology.* London: Luzac and Company.

Bach, B., Kerracher, N., Hall, K. W., Carpendale, S., Kennedy, J., & Henry Riche, N. (2016). Telling stories about dynamic networks with graph comics. In *Proceedings of the 2016 CHI Conference on Human Factors in Computing Systems.* ACM, pp. 3670–3682.

Barker, R. G. (1963). The steam of behavior as an empirical problem. In R. G. Barker (Editor), *The stream of behavior.* New York: Appleton-Century Crofts, pp. 1–22.

Barker, R. G., & Wright, H. F. (1954). *Midwest and its children: the psychological ecology of an American town.* Evanston, IL: Row, Peterson.

Bender, J., & Marrinan, M. (2010). *The culture of diagram.* Stanford, CA: Stanford University Press.

Betrancourt, M., & Tversky, B. (2000) Effect of computer animation on users' performance: a review. *Le travail Humain, 63,* 311–330.

Bierney, S., & Betrancourt, M. (2016). Does animation enhance learning? A meta-analysis. *Computers and Education, 101,* 150–167.

Bower, G. H., Black, J. B., & Turner, T. J. (1979). Scripts in memory for text. *Cognitive Psychology, 11,* 177–220.

Bordwell, D. (1985). *Narration in the fiction film.* Madison: University of Wisconsin Press.

Bordwell, D., & Thompson, K. (2003). *Film art: an introduction.* New York: McGraw-Hill.

Brewer, W. F., & Lichtenstein, E. H. (1982). Stories are to entertain: a structural-affect theory of stories. *Journal of Pragmatics, 6*(5), 473–486.

Brown, L. (1979). *The story of maps.* New York: Dover.

Brown, G., & Yule, G. (1983). *Discourse analysis*. Cambridge: Cambridge University Press.

Bruner, J. (1987). *Actual minds, possible worlds*. Harvard: Cambridge.

Bruner, J. (2004). Life as narrative. *Social Research, 71*, 691–710.

Buehner, M. J., & Cheng, P. W. (2005). Causal learning. *The Cambridge handbook of thinking and reasoning*. Cambridge: Cambridge University Press, pp. 143–168.

Buell, L. (1980). *Literary transcendentalism*. Ithaca, NY: Cornell University Press.

Card, S. K., Mackinlay, J. D., & Shneiderman, B. (1999). *Readings in information visualization: using vision to think*. San Francisco: Morgan Kaufman.

Casasanto, D. (2009). Embodiment of abstract concepts: good and bad in right- and left-handers. *Journal of Experimental Psychology: General, 138*(3), 351.

Chafe, W. (1980). *The pear stories: cognitive. Cultural, and linguistic aspects of narrative production*. Norwood, NJ: Ablex.

Chafe, W. (1998). Things we can learn from repeated tellings of the same experience. *Narrative Inquiry, 8*(2), 269–285.

Chevalier, F., Riche, N. H., Plaisant, C., Chalbi, A., & Hurter, C. (2016). Animations 25 years later: new roles and opportunities. In *AVI 16, International Working Conference on Advanced Visual Interfaces*. ACM, pp. 280–287.

Clark, H. H. (1996). *Using language*. Cambridge: Cambridge University Press.

Cleveland, W. S. (1984). Graphs in scientific publications. *The American Statistician, 38*, 261–269.

Cleveland, W. S. (1985). *The elements of graphing data*. Monterey, CA: Wadsworth.

Cleveland, W. S. (1993). *Visualizing data*. Summit, NJ: Hobart Press.

Cutting, J. E., DeLong, J. E., & Nothelfer, C. E. (2010). Attention and the evolution of Hollywood film. *Psychological Science, 21*, 432–439.

Currie, G. (2010). *Narratives and narrators: a philosophy of stories*. Oxford: Oxford University Press.

Diderot, D., & D'Alembert, J. Editors. (1751/1772). *L'Encyclopedie*. Paris: Andre le Breton.

Donald, M. (1991). *Origins of the modern mind*. Cambridge: Harvard University Press.

Dudokovic, N., Marsh, E., & Tversky, B. (2004). Telling a story or telling it straight: the effects of entertaining versus accurate retellings on memory. *Applied Cognitive Psychology, 18*, 125–143.

Eisner, W. (2008) *Comics and sequential art*. New York: Norton.

Few, S. (2012). *Show me the numbers*. Burlingame, CA: Analytics Press.

Few, S. (2013). *Information dashboard design*. Burlingame, CA: Analytics Press.

Fry, B. (2007). *Visualizing data: exploring and explaining data with the processing environment*. Sebastopol, CA: O'Reilly Media, Inc.

Gaiman, N., & McKean, D. (1987). *Violent Cases*. London: Escape Books.

Gee, J. P. (2014). *An introduction to discourse analysis: theory and method*. London: Routledge.

Groensteen, T. (2009). *The system of comics*. Jackson, MI: University of Mississippi.

Harari, Y. N. (2014). *Sapiens: a brief history of humankind*. London: Random House.

Hard, B. M., Recchia, G., & Tversky, B. (2011). The shape of action. *Journal of Experimental Psychology: General, 140*, 586–604. doi: 10.1037/a0024310

Hard, B. M., Tversky, B., & Lang, D. (2006). Making sense of abstract events: building event schemas. *Memory and Cognition, 34*, 1221–1235.

Hassig, R. (2001). *Time, history, and belief in Aztec and Colonial Mexico.* Austin: University of Texas.

Heer, J., & Robertson, G. (2007). Animated transitions in statistical data graphics. *IEEE Transactions on Visualization and Computer Graphics, 13*(6), 1240–1247.

Heer, J., & Shneiderman, B. (2012). Interactive dynamics for visual analysis. *Queue, 10*(2), 30.

Heiser, J., & Tversky, B. (2006). Arrows in comprehending and producing mechanical diagrams. *Cognitive Science, 30*, 581–592.

Hochberg, J. (2014). Organization and the Gestalt tradition. *Handbook of Perception, 1*, 179–210.

http://www.nytimes.com/2008/02/25/business/media/25asktheeditors.html.

http://www.nytimes.com/1997/01/09/world/for-third-world-water-is-still-a-deadly-drink.html.

Hullman, J., & Diakopoulos, N. (2011). Visualization rhetoric: framing effects in narrative visualization. *IEEE Transactions on Visualization and Computer Graphics, 17*(12), pp. 2231–2240.

Hullman, J., Drucker, S., Riche, N., Lee, B., Fisher, D., & Adar, E. (2013). A deeper understanding of sequence in narrative visualization. *IEEE Transactions on Visualization and Computer Graphics, 19*, 2406–2415. doi: 10.1109/TVCG.2013.119.

James, W. (1890/1981). *The principles of psychology.* Cambridge: Harvard University Press.

Kandinsky, W. (1926/2013). *Point and line to plane.* Mansfield Centre, CT: Marino.

Kessell, A. M., & Tversky, B. (2011). Visualizing space, time, and agents: production, performance, and preference. *Cognitive Processing, 12*, 43–52.

Klee, P. (1953). *Pedagogical sketchbook.* London: Faber & Faber.

Kosslyn, S. M. (2006). *Graph design for the eye and mind.*: London:OUP.

Kurby, C. A., & Zacks, J. M. (2007). Segmentation in the perception and memory of events. *Trends in Cognitive Science, 12*, 72–79.

Larkin, J. H., & Simon, H. A. (1987). Why a diagram is (sometimes) worth ten thousand words. *Cognitive science, 11*(1), 65–100.

Madden, M. (2005). *99 Ways to tell a story: exercises in style.* New York: Chamberlain Bros.

Magliano, J. P., Miller, J., & Zwaan, R. A. (2001). Indexing space and time in film understanding. *Applied Cognitive Psychology, 15*, 533–545.

Mandler, J. M. (1982). Recent research on story grammars. *Advances in Psychology, 9*, 207–218.

Marsh, E., & Tversky, B. (2004). Spinning the stories of our lives. *Applied Cognitive Psychology, 18*, 491–503.

Marsh, E. J., Tversky, B., & Hutson, M. (2005). How eyewitnesses talk about events: implications for memory. *Applied Cognitive Psychology, 19*, 1–14.

McCloud, S. (1994). *Understanding comics.* New York: Harper Collins.

McPhee, J. (2013). Structure. *The New Yorker*, January 14, pp. 46–55.

McPhee, J. (2015). Omission. *The New Yorker*, September 14.

Miyamoto, Y., Nisbett, R. E., & Masuda, T. (2006). Culture and the physical environment holistic versus analytic perceptual affordances. *Psychological Science, 17*(2), 113–119.

Morris, M. W., & Murphy, G. L. (1990). Converging operations on a basic level in event taxonomies. *Memory & Cognition, 18*, 407–418.

Munzner, T. (2014). *Visualization analysis and design.* New York: A K Peters/CRC Press.

Murch, W. (2001). *In the blink of an eye: a perspective on film editing*, 2nd ed. Los Angeles: Silman-James Press.

Newtson, D. (1973). Attribution and the unit of perception of ongoing behavior. *Journal of Personality and Social Psychology, 28*, 28–38. doi:10.1037/h0035584.

Newtson, D., & Engquist, G. (1976). The perceptual organization of ongoing behavior. *Journal of Experimental Social Psychology, 12*, 436–450. doi:10.1016/0022–1031(76)90076-7.

Newtson, D., Engquist, G., & Bois, J. (1977). The objective basis of behavior units. *Journal of Personality and Social Psychology, 35*, 847–862. doi:10.1037/0022–3514.35.12.847.

Newtson, D., Hairfield, J., Bloomingdale, J., & Cu tino, S. (1987). The structure of action and interaction. *Social Cognition, 5*, 191–237.

Neurath, O. (1936). *International picture language: the first rules of isotype.* London: Kegan Paul, Trench, Trubner & Co., Ltd.

Nisbett, R. E., & Masuda, T. (2003). Culture and point of view. *Proceedings of the National Academy of Sciences, 100*(19), 11163–11170.

Norman, D. A. (2013). *The design of everyday things.* New York: Basic Books.

Ottaviani, J. (2011). *Feynman.* New York: First second.

Pickover, C. (2009). *The math book.* New York: Sterling.

Prince, G. (2003). *A dictionary of narratology* (Revised). Lincoln, NE: University of Nebraska Press.

Queneau, R. (1947/1981). *Exercises in style.* New York: New Directions.

Ronen, R. (1990). Paradigm shift in plot models: an outline of the history of narratology. *Poetics Today, 11*, 817–842.

Rosch, E. (1978). Principles of categorization. In E. Rosch & B. B. Lloyd (Editors), *Cognition and categorization.* Hillsdale, NJ: Erlbaum, pp. 27–48.

Rumelhart, D. E. (1975). Notes on a schema for stories. In D. G. Bobrow & A. Collins (Editors), *Representation and understanding: studies in cognitive science.* New York: Academic Press, pp. 211–237.

Schmandt-Besserat, D. (1992). *Before writing, volume 1: from counting to cuneiform.* Austin: University of Texas Press.

Schubert, T., & Maass, A. (Editors) (2011). *Spatial schemas in social thought.* Berlin: Mouton de Gruyter.

Segal, A., Tversky, B., & Black, J. B. (2014). Conceptually congruent actions can promote thought. *Journal of Research in Memory and Applied Cognition, 3*, 124–130. dx.doi.org/10.1016/j.jarmac.2014.06.004.

Segal, E., & Heer, J. (2010). Narrative visualization: telling stories with data. *IEEE Transactions on visualization and Computer Graphics*, *16*, 1139–1148.

Sharer, R. J., & Traxler, L. P. (2006). *The ancient Maya*. Stanford, CA: Stanford University Press.

Shneiderman, B. (1996). The eyes have it: a task by data type taxonomy for information visualizations. In *Proceedings of the IEEE Symposium on Visual Languages*. IEEE Computer Society Press, Washington, pp. 336–343.

Sims, N. (1984). *The literary journalists*. New York: Ballantine.

Sims, N., & Kramer, M. (1995). *Literary journalism*. New York: Ballantine.

Small, J. P. (2015). *Wax tablets of the mind: cognitive studies of memory and literacy in classical antiquity*. New York: Routledge.

Smith, G. M. (2003). *Film structure and the emotion system*. Cambridge: New York: Cambridge University Press.

Smith, W. S., & Simpson, W.K. (1998). *The art and architecture of ancient Egypt*. New Haven: Yale University Press.

Speer, N. K., Reynolds, J. R., Swallow, K. M., & Zacks, J. M. (2009). Reading stories activates neural representations of perceptual and motor experiences. *Psychological Science*, *20*, 989–999.

Speer, N. K., Zacks, J. M., & Reynolds, J. R. (2007). Human brain activity time-locked to narrative event boundaries. *Psychological Science*, *18*, 449–455.

Spence, I. (2006). William Playfair and the psychology of graphs. *JSM Proceedings, American Statistical Association*. Alexandria, VA, pp. 2426–2436.

Spiegelman, A. (2013). *Co-mix: a retrospective of comics, graphics, and scraps*. Montreal: Drawn & Quarterly.

Talmy, L. (1983). How language structures space. In H. L. Pick & L. P. Acredolo (Editors), *Spatial orientation: theory, research and application*. New York: Plenum Press, pp. 225–282.

Talmy, L. (1988). Force dynamics in language and cognition. *Cognitive science*, *12*, 49–100.

Tan, E. S. (1996). *Emotion and the structure of narrative film*. Hillsdale, NJ: Lawrence Erlbaum Associates.

Taylor, H. A., & Tversky, B. (1992). Descriptions and depictions of environments. *Memory and Cognition, 20*, 483–496.

Trabasso, T., Stein, N. L., & Johnson, L. R. (1981). Children's knowledge of events: a causal analysis of story structure. *Psychology of learning and motivation*, *15*, 237–282.

Tufte, E. R. (1983). *The visual display of quantitative information*. Cheshire, CT: Graphics Press.

Tufte, E. R. (1990). *Envisioning information*. Cheshire, CT: Graphics Press.

Tufte, E. R. (1997). *Visual explanations*. Cheshire, CT: Graphics Press.

Tversky, B. (2011a). Spatial thought, social thought. In T. Schubert & A. Maass (Editors), *Spatial schemas in social thought*. Berlin: Mouton de Gruyter, pp. 75–38.

Tversky, B. (2011b). Visualizations of thought. *Topics in Cognitive Science, 3*, 499–535. doi: 10.1111/j.1756–8765.2010.01113.x.

Tversky, B. (2015). The cognitive design of tools of thought. *Review of Philosophy and Psychology. Special Issue on Pictorial and Diagrammatic Representation,* 6, 99–116 doi: 10.1007/s13164-014-0214-3.

Tversky, B., Agrawala, M., Heiser, J., Lee, P. U., Hanrahan, P., Phan, D., Stolte, C., & Daniel, M.-P. (2007). Cognitive design principles for generating visualizations. In G. Allen (Editor), *Applied spatial cognition: from research to cognitive technology.* Mahwah, NJ: Erlbaum, pp. 53–73.

Tversky, B., Corter, J. E., Yu, L., Mason, D. L., & Nickerson, J. V. (2012). Representing category and continuum: visualizing thought. In P. Rodgers, P. Cox, & B. Plimmer (Editors), *Diagrammatic representation and inference.* Berlin: Springer, pp. 23–34.

Tversky, B., & Chou, T. (2010). Depicting action: language and culture in visual narratives. Unpublished manuscript.

Tversky, B., & Chow, T. (2017). Language and culture in visual narratives. *Cognitive Semantics, 10*(2), 77–89. https://doi.org/10.1515/cogsem-2017-0008.

Tversky, B, Gao, J., Corter, J. E., Tanaka, Y., & Nickerson, J.V. (2016). People, place, and time: inferences from diagrams. In M. Jamnik & Y. Uesaka (Editors), New York: Springer.

Tversky, B. Heiser, J., & Morrison, J. (2013). Space, time, and story. In B. H. Ross (Editor), *The psychology of learning and motivation.* Oxford: Elsevier, pp. 47–76.

Tversky, B., & Kessell, A. M. (2014). Thinking in action. Special issue on Diagrammatic Reasoning. *Pragmatics and Cognition, 22,* 206–223. doi 10.175/pc22.2.03tve.

Tversky, B., Kugelmass, S., & Winter, A. (1991). Cross-cultural and developmental trends in graphic productions. *Cognitive Psychology, 23,* 515–557.

Tversky, B., Morrison, J. B., & Betrancourt, M. (2002). Animation: can it facilitate? *International Journal of Human Computer Studies. International Journal of Human Computer Studies, 57,* 247–262.

Tversky, B., & Suwa, M. (2009). Thinking with sketches. In A. B. Markman & K. L. Wood (Editors), *Tools for innovation.* Oxford: Oxford University Press, pp. 75–84.

Tversky, B., & Zacks, J. M. (2013). Event perception. In D. Riesberg (Editor), *Oxford handbook of cognitive psychology.* Oxford: Oxford, pp. 83–94.

Tversky, B., Zacks, J. M., & Hard, B. M. (2008). The structure of experience. In T. Shipley & J. M. Zacks (Editors), *Understanding events.* Oxford: Oxford University, pp. 436–464.

Van Dijk, T. A. (1993). Principles of critical discourse analysis. *Discourse & society, 4*(2), 249–283.

Van Dijk, T. A. (2001). Critical discourse analysis. In *The handbook of discourse analysis,* New York: Blackwell Publishers, pp. 349–371.

Wang, A-L. (2011). *Culture and math visualization: comparing American and Chinese math images.* Unpublished MA, thesis. Columbia Teachers College.

Ware, C. (2013). *Information visualization: perception for design.* Waltham, MA: Morgan Kaufmann.

Wiesner, D. (2001). *The Three Pigs.* New York: Clarion.

Winterowd, W. R. (1990). *The rhetoric of the "other" literature*. Carbondale: SUI Press.

Won, J. L. (2011). *Visual representations in children's picture books across cultures and languages*. Unpublished MA, thesis. Columbia Teachers College.

Zacks, J. M. (2015). *Flicker: This is your brain on movies*. Oxford: Oxford University Press.

Zacks, J. M., Braver, T. S., Sheridan, M. A., Donaldson, D. I., Snyder, A. Z., Ollinger, J. M., Buckner, R. L., & Raichle, M. E. (2001). Human brain activity timelocked to perceptual event boundaries. *Nature Neuroscience*, *4*, 651–655.

Zacks, J. M., Kumar, S., Abrams, R. A., & Mehta, R. (2009). Using movements and intentions to understand human activity. *Cognition*, *112*, 201–216.

Zacks, J. M., Kurby, C. A., Eisenberg, M. L., & Haroutunian, N. (2011). Prediction error associated with the perceptual segmentation of naturalistic events. *Journal of Cognitive Neuroscience*, *23*(12), 4057–4066.

Zacks, J. M., Speer, N. K., & Reynolds, J. R. (2009). Segmentation in reading and film comprehension. *Journal of Experimental Psychology: General*, *138*, 307–327.

Zacks, J. M., Speer, N. K., Swallow, K. M., Braver, T. S., & Reynolds, J. R. (2007). Event perception: a mind/brain perspective. *Psychological Bulletin*, *133*, 272–293.

Zacks, J. M., Speer, N. K., Swallow, K. M., & Maley, C. J. (2010). The brain's cutting-room floor: segmentation of narrative cinema. *Frontiers in Human Neuroscience*, *4*, 1–15.

Zacks, J., & Tversky, B. (1999). Bars and lines: a study of graphic communication. *Memory and Cognition*, *27*, 1073–1079.

Zacks, J., & Tversky, B. (2001). Event structure in perception and conception. *Psychological Bulletin*, *127*, 3–21.

Zacks, J. M., Tversky, B., & Iyer, G. (2001). Perceiving, remembering, and communicating structure in events. *Journal of Experimental Psychology: General*, *130*, 29–58.

Zhang, F. (2015). *Math visualizations across cultures: comparing Chinese and American math images*. Unpublished MA, thesis, Columbia Teachers College.

# CHAPTER 3

# Exploration and Explanation in Data-Driven Storytelling

Alice Thudt and Jagoda Walny

*University of Calgary*

Theresia Gschwandtner

*Vienna University of Technology*

Jason Dykes

*City University, London*

John Stasko

*Georgia Institute of Technology*

## CONTENTS

## INTRODUCTION

In visualization, it may seem that exploration and explanation are opposite ends of a spectrum: that increasing the focus on explanation and communication necessarily reduces the ability for exploration. However, data-driven stories challenge this notion by combining exploration and explanation in creative new ways.

Exploration provides readers with the power to find their own story in a set of data, while explanations communicate an author's narrative about the data. Exploratory facets are useful for data analysis and personalized navigation. Explanatory facets, on the other hand, are essential for communicating information and viewpoints about data and establishing emphasis and story lines for the reader. The purpose of visual data-driven stories is to communicate a narrative and present information based on data, so explanatory elements are a defining aspect of visual stories.

The notion of considering exploration and explanation as opposites was influential in early work in interactive cartography [1] and is seen in data journalism [2], where the journalist's goal is to provide a narrative through and supported by data. However, this distinction has collapsed as interactive interfaces have become ubiquitous. As digital technologies develop and widen the opportunities for interactive data interfaces, increasing numbers of graphics concurrently have both exploratory and explanatory characteristics. This form of visual storytelling offers new possibilities for design, engagement, discovery, and narrative devices [3,4]. Developing practice in data-driven storytelling has established compelling examples that show how powerful exploratory explanations and explanatory explorations can be (see Figure 3.1(a)–3.1(c)).

As visualization researchers, we see value in providing exploratory power to readers of data-driven stories. While explanation is a powerful way to provide a narrative for readers and orient them within large and complicated issues and data sets, exploration enables readers to make their own inquiries, personalize their reading experience, and get a feeling for the limits and the shape of the data. Providing exploratory power can also be a way to communicate complexities in the data and mitigate some of the biases inherent in providing a narrative.

We argue that making a visualization more exploratory does not necessarily take away from its explanatory value and vice versa. As shown in Figure 3.2, we view explanation and exploration as two complementary capacities of visualizations that can be successfully integrated in different ways. The "Film Dialogue" story [5] (see Figure 3.1(a)–3.1(c)) shows that this is indeed possible. The visualizations and accompanying text provide a clear narrative and message, i.e., that men dominate movie roles and dialogues. Meanwhile, the visualizations enable the reader to investigate individual films more deeply. These visualizations are interactive, show large amounts of individual data points, and are presented using multiple different types of views. Recognizing that exploration can be added to a story such as this without taking away from its explanatory power can push us to imagine more sophisticated combinations of exploration and explanation that can simultaneously communicate coherent narratives while empowering readers to gain a fuller understanding of the supporting data.

Data-driven stories that are highly exploratory and highly explanatory at the same time are still difficult to find. However, as we will show in this chapter, by breaking down exploration and explanation into more manageable dimensions, it is possible to imagine such stories. As we discuss exploration and explanation, we invite readers to consider the explanatory and exploratory aspects of the data-driven stories they have seen and to imagine how one could increase exploration and explanation simultaneously in the examples we show.

In this chapter, we document interesting examples that show how exploration and explanation interact and complement one another in the evolving landscape of data-driven storytelling. We describe the

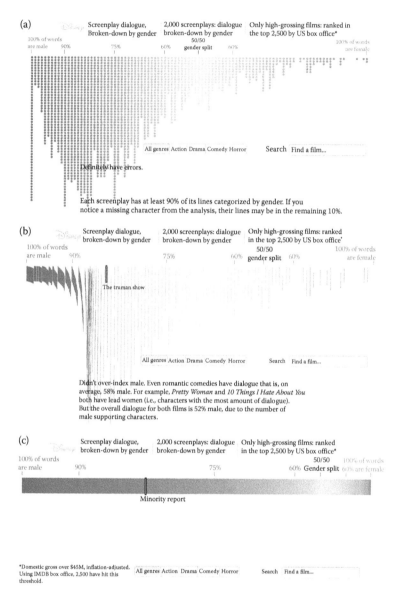

FIGURE 3.1 "Film Dialogue." (a) "Film Dialogue" is a narrative comparing female to male dialogue in 2000 screenplays of popular films. (b) When scrolling through the explanatory text, animated transitions link each view to the next. (c) The story's main message that women have far less dialogue than men is clearly communicated by the views and the accompanying text. Yet, the reader can explore each view interactively to investigate individual films. Thus, explanation and exploration complement one another to form an engaging story. (Source: Polygraph. http://polygraph.cool/films/. Used with permission. [5])

*(i) Exploration and explanation as opposites*

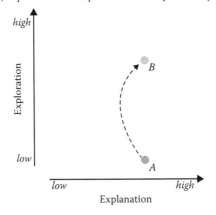

*(ii) Exploration and explanation as complementary*

FIGURE 3.2  Considering exploration and explanation as opposites (i) implies that when an exploratory visualization (A) becomes more explanatory (B), it also becomes less exploratory. In this chapter we take the position that the two can also be seen as complementary (ii) in that a visualization (A) can be made more exploratory (B) while maintaining its explanatory capacity.

exploration-explanation space in light of recent visualization research and our experience in exploratory visualizations. We identify three dimensions—view, focus, and sequence—along which exploration and explanation can vary. By viewing data-driven stories through the perspective of these dimensions, we can illustrate a variety of ways in which exploration and explanation can be integrated, and we can identify interesting features of the exploration-explanation landscape. Our hope is that this view will inspire designers to populate this landscape in new ways, that it will help developers to produce enabling technologies that extend it, and that researchers will find intriguing research questions about the relationship of exploration and explanation to communication. As all of these groups work to improve and expand the experience of reading data-driven stories, our perspective of exploration and explanation as being complementary rather than opposites draws attention to a space for designing graphics that are both highly explanatory and exploratory, hence pushing the boundaries of what is possible in data-driven storytelling.

## CHARACTERIZING EXPLORATION AND EXPLANATION IN VISUAL DATA-DRIVEN STORIES

Exploratory facets of data-driven stories, to which we refer collectively as *exploration*, are those facets that provide readers with the flexibility to ask a variety of questions of the data, to personalize their experience of the story based on their own interests, and to view the data from different perspectives.

Explanatory facets of data-driven stories, to which we refer collectively as *explanation*, involve an author's deliberate communication of context or interpretation relevant to the data or its presentation. The author's message can be communicated directly (e.g., through text or audio) or indirectly through the visualization.

### Characteristics of Exploration

Exploration is characterized by access to additional data, and the functionality or forms of the representation. Adding exploratory facets allows authors to create stories that can be personalized according to readers' knowledge, interests, and experience as they develop.

Exploration can be supported by allowing readers to choose among a large set of different preselected representations, vary the visual variables used in these representations, and change the focus by selecting particular data attributes. Including features for personalizing a story to the readers' context is another way to let readers find their own narrative (for an example, see the Neighbourhood Statistics Quiz of the UK government described in the section entitled "Flexibility in Choosing the Focus").

Even static representations can have exploratory characteristics. Consider, for instance, the network visualization of muesli ingredients [6] shown in Figure 3.3. It allows the viewer to ask multiple questions by looking at different items in the graphic. For example, a person could be interested in finding common cereal-fruit combinations or in assessing how the most popular ingredients are mixed together. The graphic shows the complexity of a data set with room for individual interpretations, while providing little explanation, narrative structure, or starting points for such explorations.

Many exploratory visualizations, however, provide interactive capabilities, especially when larger data sets are presented. Interaction allows the reader to investigate different aspects of the data that may not be initially

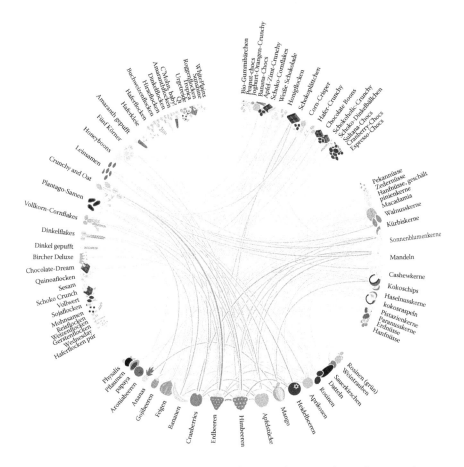

FIGURE 3.3  Static representations can have exploratory facets by providing a large enough amount of complex information that can be explored from different angles. (Source: Stefaner, M. http://truth-and-beauty.net/projects/muesli-ingredient-network. Used with permission. [6])

evident. It is the most common way to support various reader intents, including selection, filtering, abstraction, elaboration, and exploration [7].

The exploratory characteristics of a data-driven story can also vary over different stages of the narrative. For instance, in the "martini glass" structure described by Segel and Heer [8], a reader is first led along a single narrative path (the "stem" of the martini glass) and is then provided with an exploratory interactive visualization through which multiple paths are possible (the "mouth" of the martini glass).

## Characteristics of Explanation

Explanation is characterized by an authorial presence that, directly or indirectly, provides a reader with interpretation, context, or other information relevant to the subject of the story beyond the raw data.

Stories with a strong explanatory component often make use of selection and visual emphasis to guide the focus of the reader's attention to the aspects that are relevant to the message of the story. For example, Figure 3.4 shows two graphics of the same data but with differing amounts of explanation. The visualization on the left shows the data with only sufficient information to interpret the graph. The visualization on the right communicates the data in a more explanatory way by adding interpretation in the headline and corresponding emphasis within the visualization to draw attention to the message. While the same subset of data is shown in both graphs, the right one clearly suggests a focus and sequence by emphasizing the data that is most relevant to the message. This form of emphasis is typical for explanatory visualizations.

However, the explanatory aspects can be introduced in much more subtle ways than in the example above, which can be seen, for instance, in *The Guardian's* visual story about China's economic slowdown [9] (see Figure 3.5). The graphic depicts how China's economic slowdown could affect other countries' economies (Figure 3.5). It uses text and the visual metaphor of dragging down to provide an interpretation of the data.

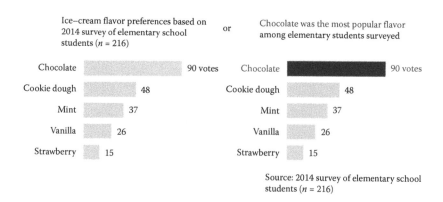

FIGURE 3.4   Two charts displaying the same data. The chart on the right has more explanatory qualities than the chart on the left; it contains added interpretation in the form of a headline and emphasis within the chart. (Source: Kemery, A., Four Storytelling Strategies. http://annkemery.com/four-storytelling-strategies/. Used with permission. [19])

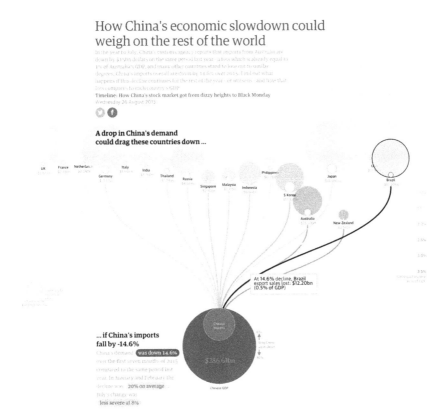

FIGURE 3.5   This interactive data-driven story contains explanatory elements, such as the headline, text, tooltips, and visual metaphor. It also contains exploratory elements: a reader can explore the effect of different values of China's economic slowdown by sliding the bottom circle, representing China, up or down. This updates the visualization of the remaining countries accordingly. (Source: *The Guardian.* http://www.theguardian.com/world/ng-interactive/2015/aug/26/china-economic-slowdown-world-imports. Used with permission. [9])

China has been specifically selected as the focus of the story here since it is the only country that can be interactively investigated, and only relationships with China are represented. In this example, further explanation is provided by adding a title and by giving interactive control over only one aspect of the visualization, i.e., the amount to which China's imports fall. The narrative is so strongly focused on China that the reader may not even think of investigating the effects of falling imports of other countries (this information is not provided in the story anyway). Hence, controlling *where* interaction is possible may be an effective way of guiding interpretation.

Besides interactive visualizations, long-form text and linear sequences are common in data-driven stories with a strong explanatory focus. Text is an effective way to communicate context and interpretation, and linear sequence is a natural companion to long-form text. Highly explanatory stories are often finite: they have a beginning, an end, and often a linear progression. There is a growing number of examples that weave text together with visualizations, as for instance, in the story on fracking in California (see Figure 3.9) in which the visualization updates as the reader scrolls through the story. There have also been efforts to automatically annotate visualizations in news articles [10].

The same techniques used for communicating an interpretation can also be used for communicating facts (as in the headline on the left bar chart in Figure 3.4), or to provide assistance with reading the visualizations or using the interactive features of the visualization.

These more-objective aspects of explanation (i.e., giving hints about where to look, annotating facts, or providing assistance with interacting with the visualization) are related to the concept of guidance [11]. However, guidance is aimed at supporting users of an interactive visualization in executing their tasks, while explanation is aimed at communicating the author's message.

Explanation can be highly visible or it can be more subtle. For example, the choice of visual encoding may influence how the message of a story comes across. However, providing explanation involves an explicit editorial process: the author deliberately chooses what to show and how to show it in order to support and communicate the story. It is author-driven and focuses on communicating specific information to the reader.

## DIMENSIONS OF DATA-DRIVEN VISUAL STORIES

Data-driven visual stories typically contain both explanatory and exploratory facets that can be integrated with each other to differing degrees. To illuminate the ways in which exploration and explanation can be varied within a data-driven story, we distinguish between two aspects of these stories: the *flexibility* they provide to the reader and the *interpretation* suggested by the author of the story. We can consider flexibility and interpretation individually in terms of three dimensions:

The *view*: The ways in which data is shown to the reader, including the available encodings and any transformations made on the data (such as normalization or aggregation).

The *focus*: The subject of the story, the particular set or subset of data shown, and the aspects of data that are shown.

The *sequence*: The order in which the information in the story can be viewed.

How flexibility and interpretation are provided on these three levels determines the degree to which a visualization supports exploration and explanation. Flexibility is most strongly related to exploration, while interpretation is most strongly related to explanation. In this section, we discuss how flexibility and interpretation can be provided through these three components, and we discuss exploratory and explanatory capabilities of data-driven visual stories in relation to these dimensions.

## Flexibility

A visualization that supports some degree of flexibility gives the viewer control over the view, focus, or sequence of the story. The flexibility that a visualization can provide is fundamentally affected by the chosen view and the complexity of the graphic. Even a static graphic with a complex view, such as the Muesli Chart in Figure 3.3, can provide flexibility in terms of the focus (the reader can choose the subset of information to examine) and the sequence (there are multiple paths to reading the graphic).

### Flexibility in the View

A visualization can provide flexibility in the view by allowing readers to control the mapping directly or choose from a number of alternative visual mappings. In visual storytelling, alternative visual mappings are frequently provided, for instance through separate views, coordinated views, or animated transitions. Techniques that allow the reader to control the visual mapping directly are less common, but they provide a high degree of flexibility. Flexibility can also be provided by allowing readers to manipulate the way the data is arranged in the visualization, for example, data aggregation or normalization (for a more thorough discussion of data specification aspects, see Chapter 10, entitled "Ethics in Data-Driven Visual Storytelling").

Below, we describe some examples of varying degrees of flexibility in view.

**Multiple separate views:** Alternative views of the same data set are shown separately, either side by side using the same type of visualization representation to show different subsets of the data (i.e., small multiples [12]) or in different parts of the text to highlight different aspects or perspectives on the data. An example of using separate views to highlight different aspects of the data is the visualization of Boston's Massachusetts Bay Transit Authority (MBTA) data (Figure 3.6). It shows adjacent views: a subway map on the left displays how many people enter and exit each station per day and a bar chart with integrated heatmaps on the right provides more information on the passenger numbers during different times of the day.

**Multiple coordinated views:** The same data is presented in multiple coordinated views that allow readers to make connections across these views, and thus, across different aspects of the data set. Such visualizations typically use a technique known as "brushing and linking." Another example of using coordinated views to give readers freedom is "A World of Terror" [13] (Figure 3.7). It shows an interlinked matrix of charts that can be rearranged and filtered.

**Animated Transitions between Views:** This method gives readers the opportunity to fluidly transition from one representation to another. The "Film Dialogue" visual story (Figure 3.1) makes use of this method. It

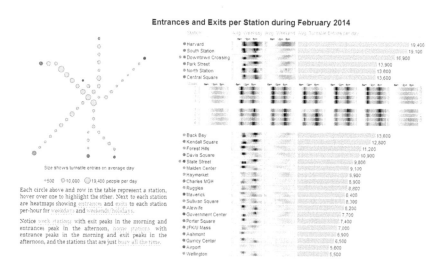

FIGURE 3.6 This graphic juxtaposes a subway map, heatmaps, bars, and text to provide a flexibility in the view. It further supports flexibility of focus by using a variety of small graphics that reveal different aspects of the data. (Source: Barry, M. and Card, B., Visualizing MBTA Data. http://mbtaviz.github.io/. Used with permission. [14])

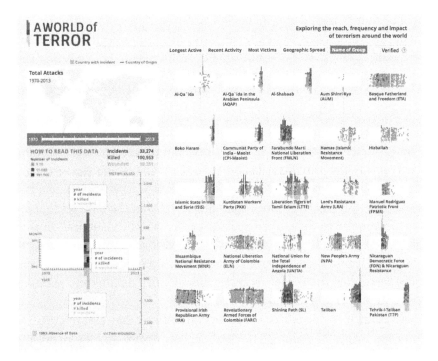

FIGURE 3.7   The multiple coordinated views shown here allow the reader to explore connections through brushing and linking. (Source: Periscopic, A World of Terror. http://terror.periscopic.com/. Used with permission. [13])

shows charts that morph as the reader scrolls through the story linking the currently shown plot to the surrounding text. Showing the transitions between different views can help to both understand different perspectives of the same data and make the relations between these perspectives easier to follow. This technique of transitioning between visualizations to match and accompany the current position in text is a recent trend in visual storytelling sometimes referred to as "scrollytelling."

**Reader-driven visual encoding:** A reader-driven visual encoding allows readers to manipulate visual variables, such as color, size, or spatial arrangement of data items, directly. For instance, in a visualization presenting data through a scatterplot, the reader may have the freedom to choose which data dimension to map on the $x$-axis, which data dimension to map on the $y$-axis, and which dimensions to portray through glyph size, shape, and color. Although not commonly adopted in visual storytelling so far, this technique provides the highest degree of flexibility in the view. One reason for the low adoption of this technique in visual stories might be that its high degree of flexibility can easily lead to confusion and unfruitful exploration. Therefore, and also to communicate the intended story, authors of visual stories often prefer to limit flexibility and add focus and sequence to increase levels of explanation.

### Flexibility in Choosing the Focus

Another way to support flexibility in a visualization is to allow readers to view different subsets of the data or decide which attributes to show. A prerequisite for providing flexibility in the focus is that the visualization offers access to a sufficient amount of data so that the readers can choose their own focus by investigating the items in the data set that are most meaningful to them. As discussed earlier, a static visualization with high visual complexity allows the reader to investigate different subsets in detail. More commonly, however, visualizations provide interaction techniques that allow the reader to change focus, such as *filtering, selecting, zooming and panning, and drill down.*

For instance, the visualization of Boston's MBTA data [14] (Figure 3.6) allows the reader to focus on various subsets of the data by providing many different visualizations on various aspects of Boston's subway system in combination with interactive focus selection.

Another example of letting readers select the focus of the story is the Neighbourhood Statistics Quiz of the Office for National Statistics in the UK [15] (Figure 3.8), which enables personalization. In the very first step, it asks the readers to enter their postal code and subsequently lets them guess demographic facts about their area while giving immediate visual feedback. Thus, while a rigid set of demographic facts are reported, a very personal data-driven story unfolds in visual form as the extent to which the reader knows an area that is important to them is revealed along with their knowledge gaps and biases.

### Flexibility in Choosing the Sequence

Narrative is all about order, but interactive stories can be designed in flexible ways that allow readers to determine order or aspects of order [16]. While static visualizations do not limit the number of viewing sequences, they may overwhelm the reader with huge amounts of data all presented simultaneously. Conventional videos, however, can only be viewed in one sequence. Besides these two extremes, interactive visualizations provide techniques to support different degrees of flexibility. For instance, one-dimensional flexibility is provided by interactions such as scrolling

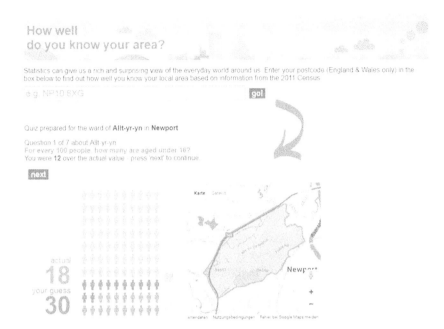

FIGURE 3.8  This website allows readers to focus on a subset of census data for a particular location of interest by inputting a postal code. (Source: UK Office for National Statistics. http://www.neighbourhood.statistics.gov.uk. Used with permission. [15])

forwards and backwards, or by "next" and "back" buttons. One example of this can be found in the graphic "California's Getting Fracked" [17] (Figure 3.9), where the visualization on the map of California is updated as the reader scrolls or steps through the story.

Two-dimensional flexibility can be provided, for instance, by panning and zooming on a map.

Many data-driven stories with strong narrative elements limit the flexibility of the viewing sequence of the presented information [18]. Such stories frequently use step-by-step (i.e., "stepper") or scrolling (i.e., "scrolly-telling") interface controls that update a story as the reader scrolls through the presentation. More flexible methods that allow a reader to explore freely typically permit the reader to jump to different visualizations in any order.

## Interpretation

Interpretation refers to any kind of meaning added by the author. Interpretations can be subjective by adding emotional aspects, persuasive,

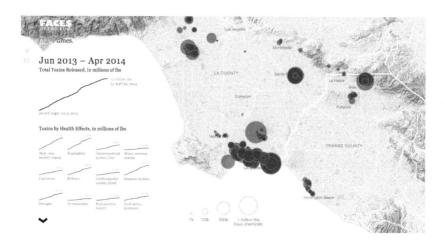

FIGURE 3.9    Readers can experience this visualization in different sequences by stepping through different points on the right or by scrolling to step through in an author-determined linear order. (Source: Flagg, A., Craig, S., and Bruno, A., California's getting fracked. http://www.facesoffracking.org/data-visualization/. Used with permission. [17])

or personal opinions. But there can also be more objective interpretations that point out important facts in the data or help the reader find interesting aspects in the chart. With *interpretation* we refer to any authorial/editorial presence in the presentation of the data. This can be conveyed through text or audio descriptions, but can go beyond a plain description of a graph and what can be seen in it.

Consider for instance the bar chart shown in Figure 3.4. The headline in Figure 3.4a says: "Ice cream flavor preferences based on 2014 survey of elementary school students" [19]. No interpretation of the data is imposed in the statement. In Figure 3.4b the caption reads: "Chocolate was the most popular flavor." This statement chooses a specific aspect of the data to point out. Additionally, this interpretation is supported through the visual encoding by highlighting the bar representing chocolate, thus guiding the focus of the reader to the intended interpretation. In this case, the interpretation is conveyed textually and through the visual representation itself. In the following section, we describe how interpretation of a visualization can be conveyed through the view, the focus, and the sequence.

### Interpretation through the View
An interpretation can be conveyed through the choice of representation, through other aspects of the visual encoding (e.g., the choice of

color or through animation), or through the choice of how the data is transformed for representation (e.g., data aggregation or data normalization). Hullman and Diakopoulos [20] describe how default views, such as showing specific comparisons or presenting the data in a specific order, prioritize some information. Moreover, the provided interaction mechanisms influence the way readers interact with the presented information, and thus, these mechanisms may lead readers to explore certain subsets of the data, and eventually, they influence the conclusions that readers draw from the visualization [16]. A powerful example of using the view to support an interpretation is "Iraq's bloody toll" [21] showing deaths in Iraq over time in a red upside down bar chart (Figure 3.10). The red supports the interpretation of the Iraq war as "bloody" and the simple decision regarding the orientation of the bars emphasizes this message. This is a basic decision, but one that has considerable effect. And while it may seem obvious—how many upside down bar charts have you ever seen?—creative redesign, even of the most fundamental principles or established norms used in our graphics, can have powerful effects.

*Interpretation through the Focus*

An interpretation can further be communicated by guiding the focus of the reader specifically to the aspects that support the story. One way of doing this is through highlighting and annotations (as in Figure 3.4) or by fading out or omitting data that is not directly relevant to the interpretation. Additionally, over the course of a story the focus can change to support the narrative. One example of varying the focus to support a story is "California's Getting Fracked" (Figure 3.9). In this example, the story starts by displaying the Los Angeles Basin area and later zooms out to reveal the entire map of California. The focus is varied over the course of the article to support the interpretation that the Los Angeles Basin is only one of the more severely polluted areas in California that is affected by fracking.

*Interpretation through the Sequence*

Many data-driven stories show visualizations in a linear sequence that supports a narrative and interpretation. Suggesting a sequence in which the data should be viewed to understand the interpretation can be done in many different ways, for example, by using steppers, scrolling, and video formats. One example that uses sequence in an effective way is

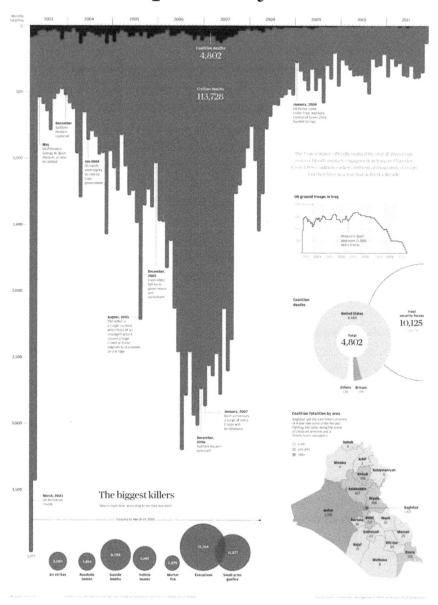

FIGURE 3.10   The visual encoding in this graphic suggests a clear interpretation of the data that fits with the author's narrative. (Source: Scarr, S. Iraq's bloody toll. http://www.scmp.com/infographics/article/1284683/iraqs-bloody-toll. Used with permission. [21])

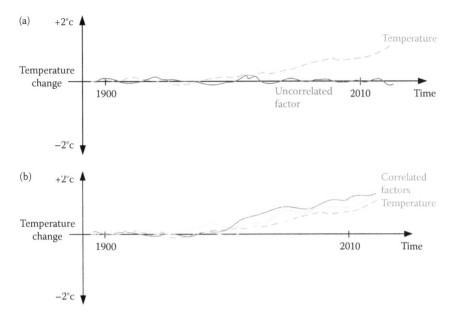

FIGURE 3.11   This data-driven story uses a linear sequence of visualizations to build a narrative. (a) Data is sequentially revealed to create suspense and support the message. First, two uncorrelated data time series are compared. (b) In the last step of the sequence, the author has placed all previously shown data within one view to support the story's conclusion. (Source: Roston, E., and Migliozzi, B. What's really warming the earth? https://www.bloomberg.com/graphics/2015-whats-warming-the-world/. Used with permission. [22])

"What's really warming the world?" [22] (Figure 3.11). Rather than showing all data at once, the story creates suspense by letting it build up step-by-step. Possible factors that have been argued to contribute to global warming are revealed over time (Figure 3.11(a)) and lead up to the conclusion that greenhouse gases are in fact the main cause of global warming (Figure 3.11(b)). In this case the sequence directly drives the interpretation of the data.

## Summary

We consider the amount of flexibility and interpretation provided in a data-driven story as being key to determining the extent to which the story supports explanatory and exploratory capabilities. An explanatory visualization provides an interpretation of the data that is deliberately communicated to readers either by choosing a particular view, focusing on specific aspects of the data, or providing a narrative sequence that supports the interpretation. Many existing storytelling devices apply here (see Chapter 6 for an explanation of storytelling devices). On top of this, an interpretation can further be provided in the text component of a data-driven story, e.g., through annotations or surrounding longer text blocks.

A visualization with exploratory aspects would typically allow for flexibility by either providing control over the choice of view, the focus, or sequence in which the data is viewed.

This flexibility can help readers to find their own interpretation in the visualization, but it is, of course, channeled in its own way through editorial decisions in the way in which flexibility is provided. Degrees of flexibility can be varied in particular parts of stories to combine the freedom of exploration with the structure of explanation.

## BENEFITS OF EXPLORATION AND EXPLANATION

If we consider both exploration and explanation as capacities that a visualization can support, authors must consider why and to what extent to include them in their data-driven visual narratives. In this section we discuss the benefits of both exploratory and explanatory visualizations.

### Benefits of Exploration

Exploratory visualizations can provide readers with a feeling of agency by allowing them to find their own interpretations of the data, sometimes in more subtle ways than the highly effective "You are Here" style used in the Neighbourhood Statistics Quiz (Figure 3.8). Viewers can go beyond predefined narratives to follow up on their own interests and questions. Another benefit of exploratory visual narratives is to provide transparency by making it possible to check some of the statements presented in a narrative through exploration of the visualization components of the story. In this way, exploratory visualizations can provide the potential to reveal biases. While sometimes a simple communication of facts can be desirable, exploratory visualizations can provide the benefit of highlighting the complexity of a subject and showing different sides in one representation. Text tends to be linear and has limited information-carrying capacity, while visualization is two-dimensional and can effectively show a skilled reader multiple relationships and nuances in trends. This can foster a balanced presentation of a topic.

## Drawbacks of Exploration

However, in order to freely explore a visualization, readers need to have a basic amount of literacy of both visualization as well as the presented topic or data. Being able to get more from an exploratory visualization than one would get from a linear narrative requires self-motivation to learn more about the topic. Data exploration takes time, attention, and cognitive investment which readers might not always be willing to provide. Crafting a tailor-made exploratory visualization for a specific data set requires significant design effort as well. Therefore, authors must carefully consider whether the readers they are targeting with their narrative will bring the motivation, skills, and time to engage with an exploratory visualization. Alternatives include limiting or going without exploratory capacities or using more generic exploratory tools.

## Benefits of Explanation

Explanatory visualizations have the benefit that the provided message or interpretation can help engage an audience with no prior knowledge of or interest in a topic. The provided guidance and suggested sequence in explanatory visualizations can make them seem more manageable and accessible to readers. The stronger editorial component in explanatory visualizations means that the author has control over the message that is being communicated. This can be beneficial since he or she can match the level of complexity to the previous knowledge of the target audience, which makes it more likely that people take something away from the visualization. Another benefit of a clear message is that the perspective that is shown might be more explicit in an explanatory visualization that can make eventual biases more obvious.

## Drawbacks of Explanation

A disadvantage of purely explanatory storytelling is that readers could get frustrated for not being able to ask and investigate their own questions about the data. Users who are increasingly used to exploring stories with interactive visualizations may miss this flexibility. In a purely explanatory story, readers may not be able to derive the substory that is actually relevant or of interest to them (as they could with personalized stories). Besides, the potential of an explanatory story to spark interest and trust heavily relies on the author's perspectives as well as narrative and presentation skills.

TABLE 1.1   Pros and Cons of Exploration from a Reader's Perspective

| Exploration Pros | Exploration Cons |
| --- | --- |
| Reader has agency of interpretation | Requires a level of visualization and data literacy |
| Ability to answer own questions | |
| Transparency: understand the source of the interpretation, potential to recognize bias | May require knowledge of the topic |
| | Requires self-motivation to learn about the topic and data |
| Can provide awareness of the complexity of a subject | Possibility of getting lost |
| | Requires time, attention, and cognitive investment from the reader |
| | An exploratory visualization tailored to a specific data set requires significant design investment |

TABLE 1.2   Pros and Cons of Explanation from a Reader's Perspective

| Explanation Pros | Explanation Cons |
| --- | --- |
| Message and points of interest are provided by the author | Readers may disagree with the perspective communicated |
| Facilitates navigating through the data | Reliant on trust in the author's perspective |
| Author does the work to engage the reader | Limited freedom to experience the story in one's own way |
| May inspire interest in unfamiliar topics | Potential omission of data that is personally important to the reader |

## SUMMARY

In this chapter we have identified several dimensions that affect the explanatory or exploratory nature of a data-driven story: the amount of flexibility (exploration) and the amount of interpretation (explanation) provided in the view, sequence, and focus of the story. This gives us a starting point for considering how to balance explanation and exploration in a story as we try to gain some traction in considering these tightly interrelated concepts.

The idea that data exploration and explanation are two ends of a continuum has been used by many and is useful for distinguishing established types of data-driven stories. However, we have presented exploration and explanation not as opposites, but as complementary characteristics. By using the dimensions of flexibility and interpretation of view, focus, and sequence, we show that exploration and explanation can be integrated to differing degrees to push the boundaries of data-driven story design (see Figure 3.12).

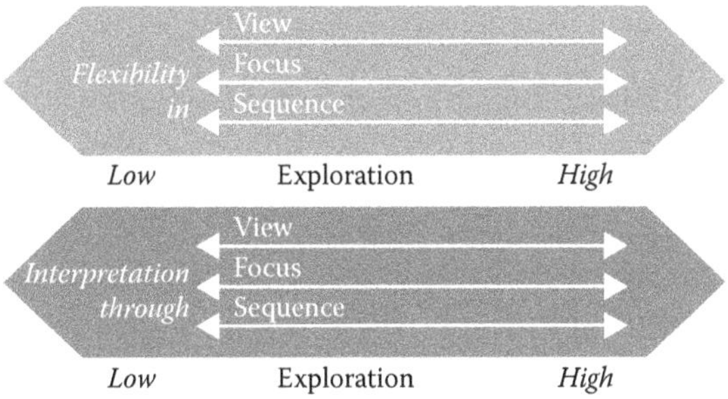

FIGURE 3.12    A data-driven story can have different levels of flexibility in its view, focus, or sequence and can convey differing levels of interpretation through its view, focus, and sequence. Manipulating a story's position along any of these dimensions can help to integrate explanatory and exploratory elements into a data-driven story.

Recognizing that exploration and explanation complement each other can inspire new ways of integrating the two beyond juxtaposing text with visualization. Even the simplest bar chart visualization provides slightly more exploratory freedom than text on its own. It is one thing to state that "Chocolate was the most popular flavor"; it is another thing to *show* that chocolate was the most popular flavor by displaying a bar chart of all flavors tested, emphasizing the bar showing chocolate, and providing context such as what other flavors were tested and how much they differed. The potential for integration goes far beyond this, however, as seen in the California fracking story in Figure 3.9, which tightly weaves explanatory text directly into the visualization.

Close balancing of exploration and explanation has the potential to achieve engaging experiences and tell persuasive stories by providing an author's context and suggested interpretation together with readers' power to personalize their experience of a story and ask their own questions. Data-driven storytelling fundamentally invites this integration of exploratory elements, a key difference from purely text-based storytelling. Thus, exploration and explanation can coexist in data-driven stories, bringing a variety of levels of freedom and interpretation into their view, focus, and sequence. Indeed, considering the various ways in which a data-driven story can support both exploration and explanation could lead to new forms of data-driven storytelling and a new understanding of what data-driven storytelling is.

## REFERENCES

1. Alan M. MacEachren. *How Maps Work: Representation, Visualization, and Design*, 1st edition. The Guilford Press, New York, 1995.
2. Alberto Cairo. *The Functional Art: An Introduction to Information Graphics and visualization*. New Riders, Berkeley, CA, 2012.
3. Robert Kosara and Jock Mackinlay. Storytelling: The next step for visualization. *Computer*, 46(5): 44–50, 2013.
4. Bongshin Lee, Nathalie Henry Riche, Petra Isenberg, and Sheelagh Carpendale. More than telling a story: A closer look at the process of transforming data into visually shared stories. *IEEE Computer Graphics and Applications*, 35(5): 84–90, 2015.
5. Hanah Anderson and Matt Daniels. Film dialogue. In: *A Polygraph Joint*, Apr 2016 (not modified). *2016 Polygraph. All Rights Reserved. http://polygraph.cool/films/ (accessed: Dec 12, 2016).
6. Moritz Stefaner. Muesli chart for mymuesli. In: *Truth & Beauty Operations* (not modified). *2012 Moritz Stefaner. All Rights Reserved. http://truth-and-beauty.net/projects/muesli-ingredient-network (accessed: Dec 12, 2016).
7. Ji Soo Yi, Youn ah Kang, John Stasko, and Julie Jacko. Toward a deeper understanding of the role of interaction in information visualization. *IEEE Transactions on Visualization and Computer Graphics*, 13(6): 1224–1231, 2007.
8. Edward Segel and Jeffrey Heer. Narrative visualization: Telling stories with data. *IEEE Transactions on Visualization and Computer Graphics*, 16(6): 1139–1148, 2010.
9. Carlo Zapponi, Sean Clarke, Helena Bengtsson, Troy Griggs, and Phillip Inman. How china's economic slowdown could weigh on the rest of the world. In: *The Guardian*, Aug 26, 2015 (not modified). *2015 The Guardian. All Rights Reserved. https://www.theguardian.com/world/ng-interactive/2015/aug/26/china-economic-slowdown-world-imports (accessed: Dec 12, 2016).
10. Jessica Hullman, Nicholas Diakopoulos, and Eytan Adar. Contextifier: Automatic generation of annotated stock visualizations. In *CHI '13 Proceedings of the SIGCHI Conference on Human Factors in Computing Systems*, ACM, New York, pp. 2707–2716, 2013.
11. Davide Ceneda, Theresia Gschwandtner, Thorsten May, Silvia Miksch, Hans-Jörg Schulz, Marc Streit, and Christian Tominski. Characterizing guidance in visual analytics. *IEEE Transactions on Visualization and Computer Graphics*, 23(1), 111–120, 2017.
12. Edward Tufte. *Visual Display of Quantitative Information*. Graphics Press, Cheshire, CT, 1983.
13. Periscopic. A world of terror. In: *Periscopic* (not modified). *2014 Periscopic. All Rights Reserved. http://terror.periscopic.com/ (accessed: Dec 12, 2016).
14. Michael Barry and Brian Card. Visualizing MBTA data. June 10, 2014 (not modified). *2014 Michael Barry and Brian Card. All Rights Reserved. http://mbtaviz.github.io/ (accessed: Dec 12, 2016).

15. UK ONS, Office for National Statistics. How well do you know your area? statistics.gov.uk (not modified). Licensed under Open Government Licence. °ONS, Office for National Statistics, UK. All Rights Reserved. http://www.neighbourhood.statistics.gov.uk/HTMLDocs/dvc147/ (accessed: Dec 12, 2016).

16. Jessica Hullman, Steven Drucker, Nathalie Henry Riche, Bongshin Lee, Danyel Fisher, and Eytan Adar. A deeper understanding of sequence in narrative visualization. *IEEE Transactions on Visualization and Computer Graphics*, 19(12): 2406–2415, 2013.

17. Anna Flagg, Sarah Craig, and Antonia Bruno. California's getting fracked. In: *Faces of Fracking* (not modified). Licensed under Creative Commons. °2014 CEL Education Fund. All Rights Reserved. http://www.facesoffracking.org/data-visualization/ (accessed: Dec 12, 2016).

18. Charles D. Stolper, Bongshin Lee, Nathalie Henry Riche, and John Stasko. Emerging and recurring data-driven storytelling techniques: Analysis of a curated collection of recent stories. Technical Report MSR-TR-2016-14, Microsoft Research, April 2016.

19. Ann K. Emery. Chocolate bar-charts. In: *Ann K. Emery Data Analysis + Visualization*, Feb 24, 2016 (not modified). °2016 Ann K. Emery. All Rights Reserved. http://annkemery.com/four-storytelling-strategies/ (accessed: Dec 12, 2016).

20. Jessica Hullman and Nick Diakopoulos. Visualization rhetoric: Framing effects in narrative visualization. *IEEE Transactions on Visualization and Computer Graphics*, 17(12): 2231–2240, 2011.

21. Simon Scarr. Iraq's bloody toll. In: *South China Morning Post*, Dec 17, 2011 (not modified). °2011 South China Morning Post Publishers Ltd. All Rights Reserved. http://www.scmp.com/infographics/article/1284683/iraqs-bloody-toll (accessed: Dec 12, 2016).

22. Eric Roston and Blacki Migliozzi. What's really warming the earth? In: *Bloomberg*, June 24, 2014 (not modified). °2014 Bloomberg L.P. All Rights Reserved. https://www.bloomberg.com/graphics/2015-whats-warming-the-world/ (accessed: Dec 12, 2016).

# CHAPTER 4

# Data-Driven Storytelling Techniques

*Analysis of a Curated Collection of Visual Stories*

Charles D. Stolper

*Southwestern University*

Bongshin Lee

*Microsoft Research*

Nathalie Henry Riche

*Microsoft Research*

John Stasko

*Georgia Institute of Technology*

## CONTENTS

## INTRODUCTION

The art of storytelling with visualization has rapidly evolved over the past few years. Advancements in web-based visualization technology and the rapid adoption of data-driven documents (D3) fostered new, dynamic data-driven storytelling on the Web (Bostock et al. 2011). Authors have introduced a variety of novel narrative visualization techniques and refined existing ones while narrative visualizations have continued to grow in popularity. Some of these stories even include innovative techniques that are new to the visualization research community. For example, "Visualizing MBTA Data" [9] tells the story of Boston's public transit using a range of interesting storytelling and visualization techniques including clever linkages between text and visualization*. We believe that the visualization community could benefit by learning about and adopting ideas from many of these stories.

In this chapter, we reflect on how authors tell stories to identify the aspects that have been under-explored by the research community and that have great potential. We identify and describe the spectrum of techniques that data-driven storytellers are using. These techniques are meant to achieve the communicative intents articulated in the narrative design patterns chapter (Chapter 5, entitled "Narrative Design Patterns for Data-Driven Storytelling"), and thus will help inform the design of future data-driven storytelling tools. In addition, we provide a curated collection of interesting examples of storytelling with data visualization and the techniques each uses. These examples should inspire and encourage authors to continue exploring ever more creative means of telling visual data stories

---

* Note: Throughout this text we use numbers in superscript brackets, e.g. [9] to refer to examples that are listed in the "Stories Analyzed" references at the end of the chapter.

TABLE 4.1    Results from Our Analysis of 45 Asynchronous, Data-Driven Stories—20 Storytelling Techniques Falling under Four Categories

| Category | Technique |
| --- | --- |
| Communicating narrative and explaining data | Textual narrative audio narration flowchart arrows labeling Text annotation on visualizations tooltips Element highlighting |
| Linking separated story elements | Linking elements through interaction Linking elements through color Linking elements through animation |
| Enhancing structure and navigation | Next/previous buttons scrolling breadcrumbs Section header buttons Menu selection timeline Geographic map |
| Providing controlled exploration | Dynamic queries Embedded exploratory visualizations Separate exploratory visualizations |

We report on a qualitative analysis of 45 stories collected from news organizations' online presences (e.g., *The New York Times*, *The Economist*, and *FiveThirtyEight*), recommended by popular visualization blogs (e.g., Visual.ly, FlowingData, Eagereyes), or featured on storytelling tools websites (e.g., Tableau Public, Infogram). To have a more focused and detailed discussion, we reviewed storytelling through Lee et al.'s lens, excluding those extreme "reader-driven" stories that lack an author-defined "plot" (Lee et al. 2015). Our set of stories ranges from more conventional narrative visualizations, such as long articles including visualizations or data-driven videos, to more experimental examples of data-driven storytelling. The results of our analysis are 20 visual, data-driven storytelling techniques falling under 4 broad categories that authors use to tell their stories (Table 4.1). In describing each of the categories, we highlight stories from the data set that exemplify the category and the associated techniques used to achieve it. We also discuss the trends we observed across these different techniques and conclude with a discussion of research opportunities, most importantly for the design of authoring tools for data-driven storytelling.

## ANALYSIS METHOD AND PROCESS

Our goal was to better understand the data-driven storytelling techniques that authors use in practice. To identify a corpus of data stories, we followed the methodologies used by previous literature (Segel and Heer 2010, Hullman and Diakopoulos 2011, Hullman et al. 2013a). We began by surveying several popular blogs' "best visualization of 2014" lists looking for those that incorporate data-driven stories (Kirk 2014a, Kirk 2014b, Skau 2014, Yao 2014, Kosara 2015). We then added a selection of popular or staff-recommended stories from the Visually community. To identify the techniques existing data-driven storytelling tools support, we included highlighted examples from a number of tools' webpages, including Tableau Story Points, Timeline JS, and Infogr.am. Finally, we supplemented this set with stories from frequent visual data journalism sources such as *The New York Times, The Washington Post, The Guardian, Bloomberg, FiveThirtyEight,* and *The Economist* to counterbalance standard practice against the novelty effect that might be present in the "best of" and "recommended" lists.

We sought to examine a broad range of stories, and we include examples that reflect recent developments in the field and that the community has found especially interesting. Our corpus is not comprehensive or representative of the field as a whole, so we will refrain from making specific quantitative ratios or judgments about a technique's prevalence. Instead, we make more general estimates on the commonality and frequency of different techniques based on the larger set of examples we considered when constructing our more-focused collection.

Unlike the broader scopes of Segel and Heer (2010) and Hullman et al. (2013a), we were specifically looking for "asynchronous, author-driven, data-driven stories." Due to the difficulty in defining the boundary between author-driven and reader-driven stories, we used a proxy condition: the stories must have some form of author-specified ordering and narrative. The ordering condition excluded the "drill-down" stories described by Segel and Heer that provide the reader complete freedom to choose the order in which to experience the story, but it included the flowchart genre they described that implies an author-chosen ordering even if the reader can ignore it. The narrative condition excluded the majority of dashboards on Tableau Public, for example, that lack any explicit authorial narrative. We did not limit stories merely to slideshows and videos, as Hullman et al. do, because this would exclude many of the scrolling-based stories that have become commonplace. Ultimately, our stories align to the category of "self-running presentations for a large audience" defined by Kosara and Mackinlay (2013).

To ensure that the pieces were data driven, we mandated that stories include at least one instance of glyph-based data visualization. This had the effect of excluding much of the output of *FiveThirtyEight*, which has a penchant for including data tables rather than visualizations in much of their work. Finally, we had many discussions among the authors regarding "list of fact" infographics (e.g., "Girl Power"* and "The Science of Happiness"†). This genre is very popular online, often consisting of a tall, vertical canvas on which a designer places a series of textual blurbs and small visuals around a theme. We differentiated these "list of fact" infographics from the infographics medium as a whole, and chose to exclude them because this specific submedium lacks authorial narrative. In the end, the resulting data set we document here consists of 45 stories, which were published after Segel and Heer's analysis—1 published in 2011, 3 in 2012, 8 in 2013, 18 in 2014, and 15 by March, 2015.

We began by analyzing each story, open tagging the narrative visualization features that we found in each. After analyzing all of the stories, we merged equivalent tags. As some tags were a higher level of abstraction than other tags, we grouped the lower-level tags under the higher-level ones, generating a set of data-driven storytelling categories and techniques. We then recoded all of the stories based on these final tags. An accompanying website‡ provides an interactive version of the table and includes direct links to each story.

## VISUALIZATION-DRIVEN STORYTELLING TECHNIQUES

In this section, we enumerate the 20 techniques we identified under four broad categories (Table 4.1). Each category is a key component of data-driven storytelling and helps differentiate these stories from more general exploratory visualization tools. While most stories in our data set employ techniques from multiple categories, the frequency of specific techniques varies strongly within each category and across the collection.

We detail each of the categories and their associated techniques. For each technique, we draw attention to specific stories from the collection that exemplify the technique. Throughout, we will employ the term "visualization" to refer to the interactive pictorial representations of data that occur in stories; "chart" when the pictorial representation is static; and the general term story "element" as a more general notion including both those pictorial representations as well as text, headers, annotations, and so on.

---

* http://www.nssfblog.com/infographic-girl-power/.
† http://www.webpagefx.com/blog/general/the-science-of-happiness/.
‡ http://www.cc.gatech.edu/gvu/ii/dds/charts/corpus.html.

Communicating Narrative and Explaining Data

As we focused our analysis on asynchronous stories without accompanying presenters, nearly every story has some form of author narration. These techniques enable authors to use the storytelling methods associated with other media, such as writing, in conjunction with data to help explain and communicate their messages.

A large number of the stories utilized the simplest variant of this category that uses long-form **textual narrative** to convey key points, interspersing visualizations throughout. Note that long-form textual narration did not always constitute paragraphs in a magazine-style article. In many slideshows, including "What the Jobs Report Really Means"[24] and "How Sunspots Control Global Weather"[17] each slide contains long-narrative text to accompany the visualization. (In this chapter, we refer to the example narrative stories via a super-scripted, bracketed number to an item in the "Stories Analyzed" section.)

Stories employed multiple forms of media for the communication of narrative and explanation of data. For example, data videos and recorded demonstrations such as "Wealth Inequality in America"[35] and "Will Saving Poor Children Lead to Overpopulation?"[20] use **audio narration**. This enables the author to more closely tie together the elements of the story (as the narrative is temporally linked to the visual elements) but in exchange for making it more difficult for a reader to experience the story at their own pace.

Moving past narrative text and audio narration, many stories include textual annotations to help communicate narrative and present data. In some cases, these annotations are **labeling** above or below a visualization or titling a slide or section (e.g., "The Yield Curve"[1]). In some cases, the annotations help guide the reader through the author's intended narrative structure. **Flowchart arrows** connect components of the story when the author's intended ordering may be unclear. For example, "Pulp Fiction in Chronological Order"[40] and "Comprehensive History of Philosophy"[41] both use flowchart arrows to help guide the reader between chronological events.

In all of the above techniques, the annotations are distinct from the charts. Conversely, many authors include on-chart annotations (i.e., **text annotation on visualizations**). These annotations can direct the reader's attention to those aspects that the author deems crucial to the story. In "342,000 Swings,"[13] the authors mark various groups of elements in the visualization such as swings per at-bat, swings per hit, and swings per home run on the visualization rather than in the textual narrative. In "The History and Future of Everything—Time,"[27] the authors mark interesting events along the animated timeline. In some cases, these events are also called out in the audio narration.

Many of the stories we analyzed provided interactive explanations in the form of **tooltips**. These tooltips vary in level of detail though. In some stories, such as "In Gaza, a Pattern of Conflict"[8] and "Bubble to Bust to Recovery"[25] (Figure 4.1), the tooltips simply show the values at a given point in the visualization. Others, such as "The World Above Us,"[45] have tooltips containing other information, in this case details about each specific satellite. One of the views in "Selfiecity"[30] has tooltips that are simply larger thumbnails of the tiny thumbnails that make up the visualization.

Other times, authors use graphical properties to perform **element highlighting**. For example, an author might wrap the highlighted elements with a shape as "Money"[33] does. Also, the author might emphasize important features while de-emphasizing the rest of a visualization (e.g., "Bubble to Bust to Recovery"[25] in Figure 4.1 and "Winning Over Virginia"[32] in Figure 4.2). Sometimes, the author may choose to only

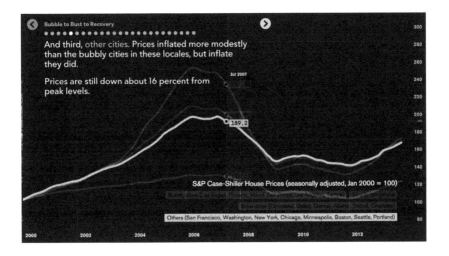

FIGURE 4.1 *Bubble to Bust to Recovery*[25] from Bloomberg. This demonstrates the use of dots as breadcrumbs, highlighting elements, tooltips, and a consistent color scheme between the textual narrative and the visualization. The legend acts as a dynamic query widget for filtering the chart.

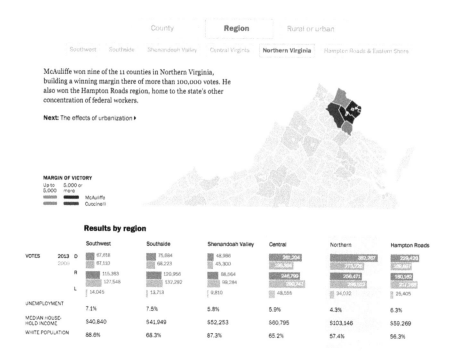

FIGURE 4.2   *Winning over Virginia* from The Washington Post. This scene demonstrates breadcrumbs with button-looking scene titles, a nested linear structure, a consistent color mapping between all of the visualization components, and author highlighting of elements in the visualizations.

color those elements that are being highlighted and maybe even add some text annotations and arrows (e.g., "The Case for a New Type of Minimum Wage"[21] in Figure 4.3 and "The Yield Curve."[1])

## Linking Separated Story Elements

Data-driven stories typically contain multiple story elements including text in various forms and visualizations or charts. Making connections between elements is crucial, particularly when data represented in visualizations is used to explain aspects of a different element. We found three fundamental types of techniques used to link different story elements.

The notion of "brushing and linking" between visualizations, where selecting items in one view highlights those same items in other views, is very common in interactive visualization systems. It is therefore noteworthy that we found only a few instances of stories utilizing the technique of **linking elements through interaction**. The only stories in our collection

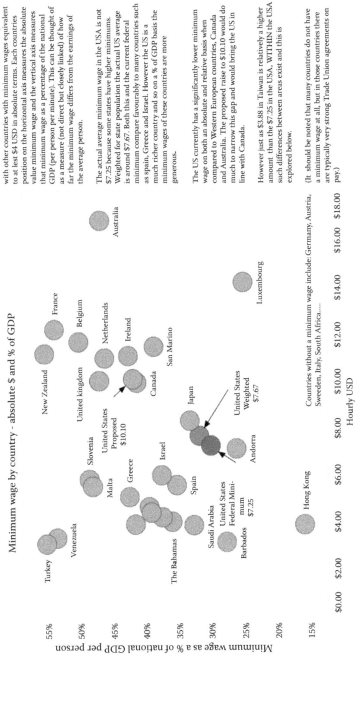

FIGURE 4.3  *The Case for a New Type of Minimum Wage*[21], a story created using Tableau Story Points. This scene provides examples of textual narrative, slides with titles as breadcrumbs, element highlighting, and textual annotation on the chart.

where interaction on one visualization is mapped to changes in another were "4th Down: When to Go for It and Why,"[12] "The Nuclear Age,"[43] "The Race Gap in America's Police Departments,"[7] and "Visualizing MBTA Data."[9]

In the first example, two different plots over the same data space (yards for a first down and position on the field) are linked such that mousing over a point in one highlights the corresponding point in the other. "The Nuclear Age"[43] consists of two visualizations: a stacked bar chart of nuclear warheads by year and country and a Gantt chart of years as a nuclear power by country. Brushing over the stacked bar chart displays the precise values for each of the country segments of the given year as labels on the Gantt chart. Clicking on a country in the Gantt chart filters the bar chart to just that country. The last story, "Visualizing MBTA Data," includes many instances of paired visualizations (usually with one being an embellished form of Boston's MBTA subway map) (see Figure 4.3).

Connections through interaction can be applied to other types of relationships. Authors incorporate brushing and linking not just between visualizations, but between the narrative text and visualizations. "Visualizing MBTA Data"[9] incorporates this technique in several ways. Hovering over certain marked phrases in the text can trigger highlighting, annotations, filtering, or opening a details panel within the visualization. The authors also use this same approach to explain how to use some of the interactive visualizations by highlighting components when the reader mouses over them in the instructions. "Road Map"[3] uses this technique to filter custom-built word clouds based on attributes described in the text. For example, a "numbered streets" filter is applied after the reader has clicked on that label embedded in the narrative.

Interaction is just one way to link separated story elements. We also found multiple examples of **linking elements through color**. Authors usually accomplish this through a consistent color mapping between attributes that appear in multiple visualizations. In "Gaza, a Pattern of Conflict,"[8] for example, all of the charts of total rocket launches are in gray, while all of the charts of deaths are in brown. In "Where We Came From and Where We Went,"[2] the authors assign each of the regions of the United States (i.e., west, south, midwest, and northeast) a color (yellow, blue, green, purple) and keep that color scheme consistent throughout the piece in all 50 of the charts as well as its breadcrumb-map (discussed in "Enhancing Structure and Navigation" section).

A useful technique is to link the narrative text and visualization through color as well. At the simplest level, this consists of sharing color-mappings between text and visualization, which often takes the form of the name of a category in the text being colored the same as the category's glyph in a visualization (e.g., "Bubble to Bust to Recovery"[25] in Figure 4.1). The link between the narrative and the visualization helps the reader discern what item in the visualization the author is referencing in the text.

Three stories in our set chose not to link multiple charts through brushing, opting instead to have a single set of glyphs that animate into the different charts that the authors use to tell the story. This **linking elements through animation** has a stronger effect on retaining the reader's context (Heer and Robertson 2007). In these examples, the transitions between the charts (or glyph arrangements) are triggered by reaching certain scrolling positions. For example, in "Green Honey"[28] when the reader scrolls to a certain point, small circles of different colors are arranged into various color spaces and then grouped by terms used in the names; "342,000 Swings"[13] uses a slight variant on this technique. The story contains a single set of glyphs, but the reader begins zoomed onto only one of the glyphs. As the reader scrolls, the story semantically zooms out and more and more of the glyphs are revealed.

## Enhancing Structure and Navigation

A key difference of storytelling from open-ended exploration is that authors impose a structure to a story and frequently provide navigation aids. Even within this notion, however, the variety of techniques authors employ demonstrates the creativity and versatility in how they encourage such structure.

A simple technique, often aligned with a slideshow-style presentation, is the use of **next/previous buttons** between slides imposing a strong linearity (e.g., Figure 4.1). An inherent linear ordering also occurs in vertically tall stories that require multiple pages of screen space. Visualizations are incorporated into a textual narrative and appear at the appropriate time (e.g., "Obama's Health Law,"[23] "In Gaza, a Pattern of Conflict,"[8] and "Beating Stanford in Fundraising"[4]). This **scrolling** technique was very common in our set and has inspired its own new genre. The technique, sometimes more explicitly called "scrollytelling" has become very popular over the last few years (Boston 2014). As the name implies, scrollytelling stories unfold as the reader scrolls; scrolling triggers changes to the visualization itself rather than simply moving the elements up or down on the screen. For example, in "342,000 Swings,"[13] while the text scrolls vertically as expected, the visualization continually zooms out as the reader scrolls down.

"Visualizing MBTA Data"[9] is an interesting twist on scrollers. Rather than having scrolling trigger any changes ("scrolljacking") a Boston subway map visualization moves down to stay on screen as the reader scrolls through a timetable visualization of each train's movement throughout a day. Because this visualization is linked with the timetable visualization (as we discussed in the "Linking Separated Story Elements" section), the reader may experience a simulated scrolljacking experience: the subway map visualization appears to be changing as the reader is scrolling when the mouse passes over the timetable visualization.

One technique common among tall stories that use a single set of glyphs (e.g., "Green Honey,"[28] "Scientific Proof that Americans are Completely Addicted to Trucks"[34] (Figure 4.4), and "342,000 Swings"[13]) are short pieces of text at various points. As the reader scrolls, these annotations scroll in the standard way while the visualizations only change at certain trigger locations. This technique provides the appropriate feedback that the scrolling is working as intended while allowing the author to tell the story.

Another technique to enhance navigation is the use of **breadcrumbs** to communicate where in the story the reader is. In most cases these breadcrumbs also provide the reader direct access to corresponding locations in the story. Authors use a variety of breadcrumbing techniques from the

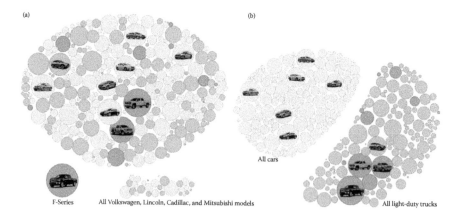

FIGURE 4.4 Two scenes from *Addicted to Trucks*[34], a story from Bloomberg. These consist of the single set of glyphs repositioned into different groups when the reader scrolls to a certain position.

trivial to the complex. A common breadcrumb style uses dots to represent each slide or scene (Figure 4.1). A number of stories use a series of **section header buttons** to convey structure and assist navigation through stories. Tableau's "8.2 Preview: Story Points" (Field 2014) has adopted this technique; the author titles each of the scenes, and these titles are placed at the top of the story to help control movement through the narrative (e.g., Figure 4.3). Other stories such as "Winning Over Virginia"[32] (Figure 4.2) use more explicit section buttons to control navigation.

In "Where We Came From and Where We Went,"[2] to handle all 50 states as separate sections and still provide direct navigation, the authors chose to use **menu selection**—from any section, the reader can select any other section from a combo-box to immediately reach that section. As the reader scrolls through the story, the menu automatically updates based on the current section (a state in the United States). This menu-for-navigation technique provides flexibility for the authors to better tell their story while retaining context for the reader. The story can have an author-chosen ordering of topics, while the menu lists the sections (in this case states) alphabetically to make it easier for the reader to find a specific section of interest. In this story, the authors also included a **geographic map** with the current state highlighted. This map is interactive, allowing the reader to jump to any of the states using it. In this manner, the map acts as a secondary navigational aid for the story.

We also observed other nonlinear navigation techniques. In "Nelson Mandela's Extraordinary Life"[44] and "Egypt in Turmoil,"[5] both made by using Timeline JS, the only visualization in the story is a **timeline** at the bottom of the display that acts as the navigational aid for the rest of the story. Each slide is represented in the timeline by either a flag (for points) or a bar (for durations). As the reader moves through the events in the story, the respective glyph is highlighted on the timeline. As with other techniques, clicking the glyph in the timeline jumps the reader to that scene in the story.

## Providing Controlled Exploration

Unlike pure exploratory visualization, narrative visualization has been characterized by limited reader interaction with visualizations. While studies have cast doubts on the traditional assumption that exploration inherently increases reader engagement (Boy et al. 2015), it is still appealing to add more flexible exploratory capabilities to stories. However, enabling full reader exploration in a story has risks. If the reader can change a visualization significantly, the data displayed may no longer be consistent with the narrative surrounding it. Furthermore, the reader may find themselves lost in the data, either from data overload or from having over-filtered. In light of this, data-driven storytelling often relies on providing exploratory elements that are controlled or constrained.

An interaction technique that has become common to constrain readers' ability to adjust visualization components is the use of **dynamic queries** (Shneiderman 1994). At end of the story, the authors of "Addicted to Trucks"[34] (Figure 4.4) allow the reader to change the color scheme and grouping attribute of one visualization independently, but constrain the choices to only four options: "car/truck," "origin," "major brand," and "gainers/losers." The dynamic query portion of "Addicted to Trucks"[34] is also an example of the **embedded exploratory visualizations** technique. This technique is used when the authors have included a more exploratory visualization, often defined by a large amount of user interaction, within the story. "Gay Rights in the US, State by State"[22] also utilizes the technique, including a large exploratory visualization at the top of the story and then discussing the details of the story as it progresses.

A final technique in the controlled exploration category is a **separate exploratory visualization** with which the reader can interact. We debated whether the inclusion of this technique violates our scope of not allowing totally open, reader-driven exploration. This technique is slightly different, however, in that the story includes a strong narrative and the authors have chosen to link to an open-ended exploratory visualization. For example, in "Selfiecity,"[30] "Oscars 2015: Does Winning Best Director Kill your Career?,"[18] "A Better Way to Find the Best Flights and Avoid the Worst Airports,"[39] and "Stop and Frisk All but Gone from New York,"[11] the authors provide a link to a separate exploratory visualization to clearly separate the open exploration from the story itself. This technique can be seen as an evolution of the "martini glass" structure identified by Segel and Heer (2010).

## DISCUSSION AND FUTURE WORK

### Composing Multiple Techniques

Through our analysis, we observed that data-driven storytelling techniques do not exist in isolation. Most stories in our set use techniques from all or many of the categories to achieve their narrative goals. For example, "In Gaza, a Pattern of Conflict"[8] uses scrolling to enhance structure and navigation, textual narrative to communicate the narrative, and color mapping to link its visualizations. In addition, it is not unusual that stories would use multiple techniques within a category. For example, "Strong Hiring Still Isn't Bringing Pay Raises"[14] includes both traditional long-form textual narrative and brief audio narration.

Authors taking advantage of scrolling navigation have used a variety of means of integrating the navigation technique with other techniques. For example, in "A Nation Divided,"[10] the authors have embedded slideshows of maps of Germany (colored by various features) within the vertical narrative. The reader advances the slides using the same scrolling navigation as the surrounding narrative text, advancing through these map slides horizontally by scrolling vertically along the page. This use of scrolljacking instead of the next and previous buttons strengthens the linearity of the story. However, there can be some confusion for the reader about the way to advance the story through the slideshows. "Selfiecity,"[30] "Where We Came From,"[2] and "Green Honey"[28] each integrate their scrolling navigation with their navigation widgets. As the reader scrolls through the stories, the story's breadcrumbs, section headers, or menu are updated to be at the reader's current position.

In addition to straight linear orderings, authors have begun to use nesting techniques. For example, "Winning over Virginia"[32] demonstrates that one slide can have subslides (such as the counties or regions in that story). In some cases these subplots might consist of a sequence of distinct visualizations, such as the map slideshows embedded within "A Nation Divided."[10] In other cases, the author might add more and more detail to a given visualization to produce narrative. In "Bubble to Bust to Recovery,"[25] for example, the author uses a sequence of slides to add one data series at a time (Figure 4.1). This growing complexity of data-driven stories illustrates how the field is maturing, and authors are utilizing the different techniques available to them to build more sophisticated visual stories.

## Opportunities for Authoring Tools

In this section, we reflect on opportunities derived from our analysis specifically for designing data-driven storytelling authoring tools. This complements the high-level research opportunities provided by Kosara and Mackinlay (2013) as well as Lee et al. (2015).

*Controlled Reader Interaction and Experience*

While there is no guarantee that interaction inherently increases reader engagement (Boy et al. 2015), authors can utilize well-designed interaction to make reading experiences more effective and fun. The main challenge we face is how to support authors to find the right balance between full exploration systems and static stories. One possible solution is to make it easier for authors to configure the degree of possible interaction with underlying data. Some of this can be achieved from supporting event-driven interaction, e.g., Ellipsis (Satyanarayan and Heer 2014) and some can be achieved from tools for mapping dynamic query widgets to elements of a story.

The story-consumption experience itself can also be reader-specific. For example, some readers might prefer a slideshow to a scroller or vice versa. Some readers might prefer static graphics with select annotations while others might be more partial to an interactive chart with tooltips. Animating charts can be fun the first time a reader experiences a story, but may become tedious on subsequent readings. In other words, readers may want to choose alternate techniques within a single category than what the author initially chose. How to leverage a reader's preferences while still remaining true to the author's story could be an interesting and valuable research direction. For more detailed discussion about broader reader preferences and capabilities, refer to the audience reception chapter (Chapter 9, "Communicating Data to an Audience").

*Smart, Dynamic, Data-Driven Annotations*

As is evidenced by the overwhelming number of example stories in both Segel and Heer's and our corpora, annotation plays a crucial role in asynchronous data storytelling as the surrogate for the storyteller. Hullman et al. (2013b) carefully studied annotation on charts and determined different types and purposes of annotations. They also created a system that can automatically produce annotated line graphs of a stock's behavior.

However, annotation is far from a "solved problem." There are three properties of any annotation: the content (e.g., text or value), the location, and the time (for the annotation to appear and disappear). Authors need to be able to control each of these aspects for each annotation. If authors provide more freedom to their readers in the course of their pieces, then there is also the risk that the annotations become inconsistent or irrelevant. Should the reader zoom or filter a view such that the data an annotation refers to is no longer displayed, then the annotation may no longer be consistent with the visualization.

In addition, authors may want to provide annotations when certain conditions are met, such as zooming deep into a view, or change the annotated text based on reader interaction. Each annotation in "The Yield Curve"[1] stays in place above the data point it references even as the reader rotates the 3D model. Ellipsis provides some initial work into this task, yet how to effectively enable better authorial control over annotations will continue to be a promising research challenge going forward (Satyanarayan and Heer 2014).

*Navigation through Scrolling*

As discussed above, scrolling has become a pervasive and powerful technique used in data-driven storytelling. Many of the most innovative stories in our collection employ scrolling in creative ways. However, only casual discussions of the merits of using scrolling as a navigational structure have been written (Bostock 2014). We need to better understand the strengths and weaknesses of using scrolling as an interaction mechanism with visualization. Furthermore, we need to incorporate scrolling into storytelling tools: either through providing authors variety in their linear structure techniques or through tools specifically designed to harness this technique.

*Effectiveness-Informed Authoring*

In our analysis, we observed the effect that tools have on the techniques that are used by authors: stories created using Tableau Story Points all had slideshows with section title breadcrumbs; Timeline JS stories were all slideshows using timeline breadcrumbing. Storytelling tools could make it easy for people to create compelling data-driven stories as well as less-effective ones. For example, excepting "Winning Over Virginia,"[32] the stories that use nested navigational structures did not indicate this explicitly in their breadcrumbs. Instead, these authors opted to treat each step as a distinct scene. It would be worthwhile to explore which of these alternatives is more effective, taking into account that there is a spectrum between explicitly presenting the nested structure and hiding it. Furthermore, a wide variety of means can express this structure, some of which authors may not even have implemented yet. Adding animation to these structures, as some authors do, further complicates matters. While animations usually help readers retain context (Heer and Robertson 2007), specifying animation often increases the complexity of both storytelling and the tools for doing so. To design a storytelling tool to encourage the effective techniques, we need to better understand their strengths, weaknesses, and ideal applications, and compare the effectiveness of the techniques within a category in a more-controlled setting.

## CONCLUSION

Data-driven storytelling continues to evolve and authors are developing new ways to support narratives with visualization. To identify the techniques that data-storytellers use, we analyzed a collection of data-driven stories, focusing on the specific techniques used in author-constructed, data-driven storytelling. We identified 20 visual storytelling techniques organized into four categories: communicating narrative and explaining data, linking separated story elements, enhancing structure and navigation, and providing controlled exploration. Through identifying techniques to be incorporated into stories, this work informs the design of future tools that enable storytellers to effectively create data-driven pieces. We also reflected on these findings and enumerated a number of possibilities for future research.

## REFERENCES

M. Bostock. How to Scroll. https://bost.ocks.org/mike/scroll/, Nov. 2014.

M. Bostock, V. Ogievetsky, and J. Heer. D3: Data-Driven Documents. *IEEE Transactions on Visualization and Computer Graphics*, 17(12): 2301–2309, 2011.

J. Boy, F. Detienne, and J.-D. Fekete. Storytelling in Information Visualizations: Does It Engage Users to Explore Data? In *Proceedings of the 33rd Annual ACM Conference on Human Factors in Computing Systems, CHI '15*, pp. 1449–1458, 2015. Seoul, Republic of Korea.

E. Fields. 8.2 Preview: Story Points. https://www.tableau.com/about/blog/2014/5/82-preview-tell-story-your-data-story-points-30761.

J. Heer and G. Robertson. Animated Transitions in Statistical Data Graphics. *IEEE Transactions on Visualization & Computer Graphics*, 13(6): 1240–1247, 2007.

J. Hullman, N. Diakopoulos, and E. Adar. Contextifier: Automatic Generation of Annotated Stock Visualizations. In *Proceedings of the ACM Conference on Human Factors in Computing Systems, CHI '13*, pp. 2707–2716, 2013b. Paris, France.

J. Hullman and N. Diakopoulos. Visualization Rhetoric: Framing Effects in Narrative Visualization. *IEEE Transactions on Visualization & Computer Graphics*, 17(12): 2231–2240, 2011.

J. Hullman, S. Drucker, N. Riche, B. Lee, D. Fisher, and E. Adar. A Deeper Understanding of Sequence in Narrative Visualization. *IEEE Transactions on Visualization and Computer Graphics*, 19(12): 2406–2415, 2013a.

A. Kirk. 10 Significant Visualisation Developments: January to June 2014. http://www.visualisingdata.com/2014/08/10-significant-visualisation-developments-january-to-june-2014/.

A. Kirk. 10 Significant Visualisation Developments: July to December 2014. http://www.visualisingdata.com/2014/12/10-significant-visualisation-developments-july-december-2014/.

R. Kosara. The State of Information Visualization, 2015. https://eagereyes.org/blog/2015/the-state-of- information-visualization-2015.

R. Kosara and J. Mackinlay. Storytelling: The Next Step for Visualization. *Computer*, 46(5): 44–50, 2013.

B. Lee, N. Henry Riche, P. Isenberg, and S. Carpendale. More than Telling a Story: Transforming Data into Visually Shared Stories. *IEEE Computer Graphics and Applications*, 35(5): 84–90, 2015.

A. Satyanarayan and J. Heer. Authoring Narrative Visualizations with Ellipsis. *Computer Graphics Forum*, 33: 361–370, 2014.

E. Segel and J. Heer. Narrative Visualization: Telling Stories with Data. *IEEE Transactions on Visualization & Computer Graphics*, 16(6): 1139–1148, 2010.

B. Shneiderman. Dynamic Queries for Visual Information Seeking. *IEEE Software*, 11(6): 70–77, 1994.

D. Skau. 14 of 2014's Best Information Designers and Animators. https://visual.ly/blog/14-2014s-best-information-designers-animators/.

N. Yau. The Best Data Visualization Projects of 2014. http://flowingdata.com/2014/12/19/the-best-data-visualization-projects-of-2014-2/.

## STORIES ANALYZED

1. G. Aisch and A. Cox. A 3-D View of a Chart That Predicts The Economic Future: The Yield Curve. http://www.nytimes.com/interactive/2015/03/19/upshot/3d-yield-curve-economic-growth.html, 2015.

2. G. Aisch, R. Gebeloff, and K. Quealy. Where We Came From and Where We Went, State by State. http://www.nytimes.com/ interactive/2014/08/13/upshot/where-people-in-each- state-were-born.html, 2014.

3. G. Aisch and B. Marsh. Road Map. http://www.nytimes.com/interactive/2015/01/29/sunday-review/road-map-home-values-street-names.html?, 2015.

4. A. Ambrose. Beating Stanford in Fundraising. https://infogr.am/beyond_crowdfunding, 2013.

5. A. America. Egypt in Turmoil. http://america.aljazeera.com/ articles/timeline-egypt-inturmoil0.html, 2013.

6. G. Arnett. The Numbers behind the Worldwide Trade in Drones. http://www.theguardian.com/news/datablog/2015/mar/16/numbers-behind-worldwide-trade-in-drones-uk-israel, 2015.

7. J. Ashkenas and H. Park. The Race Gap in America's Police Departments. http://www.nytimes.com/interactive/2014/09/03/us/the-race-gap-in-americas-police-departments.html, 2014.

8. J. Ashkenas, A. Tse, and K. Yourish. In Gaza, a Pattern of Conflict. http://www.nytimes.com/interactive/2014/07/31/world/ middleeast/in-gaza-a-pattern-of-conflict.html, 2014.

9. M. Barry and B. Card. Visualizing MBTA Data. http://mbtaviz. github.io, 2014.

10. L. Borgenheimer, P. Blickle, J. Stahnke, S. Venohr, C. Bangel, M. Young, F. Mohr, M. Biedowicz, and A. M.. A Nation Divided. http://zeit.de/feature/ german-unification-a-nation-divided, 2014.

11. M. Bostock and F. Fessenden. Stop and Frisk All but Gone from New York. http://www.nytimes.com/interactive/2014/09/19/nyregion/stop-and-frisk-is-all-but-gone-from-new- york.html, 2014.

12. B. Burke, S. Carter, J. Daniel, T. Giratikanon, and K. Quealy. 4th Down: When to Go for It and Why. http://www.nytimes.com/2014/09/05/upshot/ 4th-down-when-to-go-for-it-and-why.html, 2014.

13. S. Carter, J. Ward, and D. Waldstein. 342,000 Swings Later, Derek Jeter Calls It a Career. http://www.nytimes.com/interactive/2014/ 09/14/sports/ baseball/jeter-swings.html, 2014.

14. B. Casselman. Strong Hiring Still Isn't Bringing Pay Raises. http://fivethir-tyeight.com/datalab/strong-hiring-still-isnt-bringing-pay-raises/, 2015.

15. B. Casselman. What Wal-Mart's Pay Increase Means for the Economy. http://fivethirtyeight.com/datalab/what-wal-marts-pay-increase-means-for-the-economy/, 2015.

16. H. Enten. Boston Breaks Snowfall Record with Amazing Finish. http:// fivethirtyeight.com/datalab/boston-breaks- snowfall-record-with-an-amazing-finish/, 2015.

17. M. Francis. How Sunspots Control Global Weather. http://public.tableau-software.com/views/SunSpotsStory/SunspotsTheWeather, 2014.

18. W. Franklin, T. Griggs, and G. Arnett. Oscars 2015: Does Winning Best Director Kill Your Career? https://www.theguardian.com/film/ng-interactive/2015/ feb/20/what-it-really-means-to-win-the-oscars-best-director, 2015.

19. A. L. G. The Evolution of Israeli Politics. https://www.economist.com/ blogs/economist-explains/2015/03/economist-explains-11, 2015.

20. Gapminder Foundation. Will Saving Poor Childen Lead to Overpopulation? http: /www.gapminder.org/videos/will-saving-poor-children-lead-to-over-population/, 2014.

21. P. Gilks. The Case for a New Type of Minimum Wage. http://public. tableausoftware.com/views/MinimumWage_3/MinimumWage-StoryPointsEdition, 2014.

22. Guardian US Interactive Team. Gay Rights in the US, State by State. http:// www.theguardian.com/world/interactive/2012/may/08/gay-rights-united-states, 2012.

23. E. Klein, M. Yglesias, L. McGann, E. Barkhorn, A. Rockey, L. Williams, K. Keller, J. Posner, M. Bell, Y. Victor, R. Mark, A. Palanzi, T. Whiting, J. S. Maria, J. Fong, T. Whiting, A. Katakam, M. Thielking, A. Espuelas, E. Caswell, D. Stanfield, J. Posner, and J. Posner. Obama: The Vox Conversation. http://www.vox.com/a/barack-obama-interview-vox-conversation/obama-domestic-policy-transcript, 2014.

24. M. C. Klein. What the Jobs Report Really Means. http://www.bloomberg. com/dataview/2013-11-08, 2013.

25. M. C. Klein. Bubble to Bust to Recovery. http://www.bloomberg.com/data-view/2014-02-25/bubble-to-bust to-recovery.html, 2014.
26. M. C. Klein. How Americans Die. http://www.bloomberg.com/data-view/2014-04-17/how-americans-die.html, 2014.
27. Kurzgesagt. The History and Future of Everything—Time. https://www.youtube.com/watch?v=2XkV6IpV2Y0, 2013.
28. M. Lee. Green Honey. http://muyueh.com/greenhoney, 2013.
29. Lemonly. What Is Excessive Alcoholism Costing Our Economy? http://americanaddictioncenters.org/hidden-costs-of-alcoholism, 2015.
30. L. Manovich, M. Stephaner, M. Yazdani, D. Baur, D. Goddemeyer, A. Tifentale, N. Hochman, and J. Chow. Selfiecity. http:// selfiecity.net, 2014.
31. D. McCandless, D. Swain, M. McLean, M. Quick, and C. Miles. Byte-Sized. http://www.bbc.com/future/story/20130621-byte-sized-guide-to-data-storage, 2013.
32. T. Mellnik, K. Park, and R. Johnson. Winning over Virginia. http://www.washingtonpost.com/wp-srv/special/local/2013-elections/demographics, 2013.
33. R. Monroe. Money. http://xkcd.com/980/, 2011.
34. A. Pearce, B. Migliozzi, and D. Ingold. Scientific Proof That Americans are Completely Addicted to Trucks. http://www.bloomberg.com/ graphics/2015-auto-sales, 2015.
35. Politzane. Wealth Inequality in America. https://www.youtube.com/watch?v=QPKKQnijnsM, 2012.
36. K. Quealy and M. Sanger-Katz. Obamacare: Who Was Helped Most. http://www.nytimes.com/interactive/2014/10/29/upshot/obamacare-who-was-helped-most.html, 2014.
37. G. S., P. K., A. C. M, and L. P. Battle Scars. http://www.economist. com/blogs/graphicdetail/2014/07/daily-chart-21, 2014.
38. A. Sedghi. Zero Hour Contracts in Four Charts. http://www.theguardian.com/news/datablog/2015/feb/25/zero- hour-contracts-in-four-charts, 2015.
39. N. Silver. A Better Way to Find the Best Flights and Avoid the Worst Airports. http://fivethirtyeight.com/features/fastest-airlines-fastest-airports, 2015.
40. N. Smith. Pulp Fiction in Chronological Order. http://visual.ly/ pulp-fiction-chronological-order, 2012.
41. SuperScholar.org. Comprehensive History of Philosophy. http://super-scholar.org/comp-history-philosophy, 2015.
42. The Data Team. The Agony of Greece. http://www.economist.com/blogs/graphicdetail/2015/03/daily-chart-0, 2015.
43. The Data Team. The Nuclear Age. http://www.economist.com/blogs/graph-icdetail/2015/03/interactive-daily-chart, 2015.
44. TIME Staff. Nelson Mandela's Extraordinary Life. http://world.time.com/2013/12/05/nelson-mandelas-extraordinary-life-an-interactive-timeline, 2013.
45. D. Yanofsky and T. Fernholtz. The World above Us. http://qz.com/296941/interactive-graphic-every-active-satellite-orbiting-earth, 2014.

# CHAPTER 5

# Narrative Design Patterns for Data-Driven Storytelling

Benjamin Bach
*University of Edinburgh*

Moritz Stefaner
*Truth & Beauty*

Jeremy Boy
*NYU Polytechnic School of Engineering*

Steven Drucker
*Microsoft Research*

Lyn Bartram
*Simon Fraser University*

Jo Wood
*City, University of London*

Paolo Ciuccarelli
*Polytechnic University of Milan*

Yuri Engelhardt
*University of Twente*

# Ulrike Köppen

*Bayerischer Rundfunk*

# Barbara Tversky

*Stanford University*

## CONTENTS

## INTRODUCTION

The terms "story" and "narrative" are often used interchangeably. They also harbor different meanings in different communities (see Chapter 1, entitled "Introduction"). Here we are guided by the definitions in Gerard Genette's *The Narrative Discourse* [1], Bruner's *The Actual Mind* [2], and which are summarized in the *Media Student's Book* [3]. A **story** is defined as *all of the events in a narrative, those presented directly to an audience and those which might be inferred* [3]. In a wider sense, the story is the facts—the data—including characters, places, times, and actions important to the specific plot. A **narrative**, on the other hand, can be described as *the 'telling' of a sequence of [those] events [...] [which] shape the events, characters, arrangement of time, etc. in very particular ways so as to invite particular positions towards the story on the part of the audience* [3].

Thus, a narrative gives shape to the unfolding events in a story, with the goal of making them clear and compelling to an audience. The narrative can be linear, e.g., representing the actual order of events (the story) chronologically; or nonlinear, e.g., circular, where the beginning and end mirror themselves. Nonlinear storytelling has been used in highly acclaimed movies, such as *Pulp Fiction* and *Memento*, where the narration either jumps between scenes (*Pulp Fiction*), or tells the story backwards (*Memento*, following the main character gradually remembering his personal past) [4].

Consequently, a single story can be narrated in an endless number of ways. Beyond the temporal ordering of events, it can be told at a varying pace to create emphasis: slowing down the narrative pace can give the audience time to think about what is being "told" and to increase the importance of this moment in the story; speeding up the narrative pace can produce an overwhelming effect of information onslaught and reduce the importance of moments in the story. Arguments can be made explicit or kept implicit. The narrative can emphasize comparison and contrasting points of view; it can convey abstract concepts through the use of analogies, making them more concrete; it can address and engage the audience directly with abrupt questions, while interrupting the flow of the story; or it can draw the audience in so that it becomes a part of the story, releasing it with a call-to-action. Each variation is likely to have different effects on audiences' reactions and emotions, as well as on their understanding of the story.

Creating compelling narratives is a labor-intensive process that requires expertise, creativity, iterations, and feedback. It calls for a deep understanding and creative use of media, as well as a sense for explaining, convincing, persuading, and especially engaging audiences. There is a wealth of guidelines and best practices (e.g., [5,6]) for creating narratives in the more-established disciplines that involve storytelling, such as literature, the performing arts, comics, radio drama, theater, and cinema. However, these fall short for stories based on data that use graphic representations [7–9] (visualizations) to explain facts and develop evidence-based arguments. The goal of this chapter—in a wider sense—is to explore such guidelines for data-driven storytelling.

This chapter presents a set of narrative design patterns (*narrative patterns*) that we believe can help storytellers design data-driven narratives that rely heavily, but not exclusively, on data visualization. Narrative patterns help connect the form of the narration with the intent of a story. They are meant for journalists, Web and visualization designers, presenters, public speakers, and others trying to shape compelling data-driven stories, and engaging interactive environments. The only requirements for using narrative patterns are to have:

1. A story,

2. An idea of who your audience is, and

3. To know the effect your story and narration should have on that audience. This can include sympathy or distaste, encouragement for action, information, explanation, and so on.

It may further be useful to have an understanding of the benefits and limitations of using different media (e.g., text, numbers, images, visualization, audio, etc.), as well as the culture of the audience, as these factors usually drive the ways in which stories are told: different audiences might react very differently to the choice of media, or to the shaping of the story.

As we believe narrative patterns are best explained through examples, we have created an interactive browsable collection, available online at http://napa-cards.net (Figure 5.1). This collection is inspired by *method*

FIGURE 5.1 Three examples of narrative pattern "cards," browsable online at http://napa-cards.net.

*cards*, i.e., design cards used to stimulate creativity and facilitate ideation, that are commonly used by design practitioners and visual artists (see, e.g., IDEO cards [10], oblique strategies [11], design-with-Intent [12], creativity cards [13], service design tools [14]), or the more recent VizItCards for teaching visualization techniques [15] and data comics design patterns [16]. We firmly believe in the power of data-driven storytelling, and our experience as academics, designers, and journalists leads us to claim that narrative patterns will undoubtedly facilitate further structuring and discussion of this emerging field.

In the rest of this chapter, we describe our set of narrative patterns, and introduce a higher-level structure for grouping them. We then discuss possible combinations of patterns, using examples of data-driven narratives. After that, we detail certain aspects related to the application of patterns. We conclude with an outlook of open questions and directions of future work that results from our discussions.

## NARRATIVE PATTERNS

We define narrative patterns as follows:

*A narrative pattern is a low-level narrative device that serves a specific intent. A pattern can be used individually or in combination with others to give form to a story.*

The specific intents and possible combinations of patterns should be defined by the narrator (or storyteller). These are usually influenced by, e.g., the data, the formal setting, or the particular audience and its assumed background knowledge. The narrator may try different unique patterns and/or combinations to see how they fit the intent(s), or use the concept of patterns to analyze existing stories and reflect on their specific intent(s). Examples of narrative intents range from enlightening audiences, to evoking empathic response, to engaging them to take action, or to questioning their beliefs and behavior. The goal may be to trigger dialog, or simply to immerse people in an enjoyable experience. These are all rather traditional intents, and are common to all kinds of storytelling disciplines (e.g., cinema, theater, etc.). They are a good starting point for data-driven storytelling. Other, more specific intents for data-driven storytelling may include delivering convincing arguments backed with data, explaining a type of data, and sensitizing people to their existence and power, or simply educating an easily targeted online audience.

In the context of data-driven storytelling, narrative patterns should not be considered direct implementations of visualization or interaction techniques, nor should they be bound to any specific implementation. In most cases, they are decoupled from the final medium and visual presentation. Patterns can overlap, they can be fuzzy, and in some cases, they may only be applicable to a single type of story.

We initially identified around 40 distinct narrative patterns based on a great number of data-driven stories that we found in the literature and on the Web. Through extensive discussions, we were able to narrow down this initial number to a core of 18 patterns. While we focus mainly on these 18, we stress that any existing story can inspire new patterns, and that our collection is by no means definitive. To better understand these patterns, and to facilitate their discussion and application, we categorize them in five different groups: *argumentation, narrative flow, framing, empathy and emotion*, and *engagement*.

This typology (Figure 5.2) was derived from an initial set of open-ended questions, such as: *What types of pattern exist? What is their specific purpose? How are they employed? How can they be applied in a story? Can a story be told with a different set of patterns?* Note that individual patterns can serve multiple intents, and that our typology is not meant to be exclusive. It merely presents an initial framework to think about narrative patterns, and aims to facilitate their application. In our online collection, patterns can be looked up according to their main intent(s), and each is illustrated with an existing, published example (Figure 5.3).

## Patterns for Argumentation

Following the *Oxford Dictionary*, argumentation is *the action or process of reasoning systematically in support of an idea, action, or theory.*

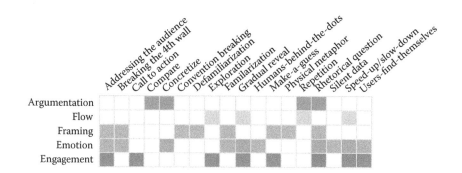

FIGURE 5.2    18 narrative patterns grouped into 5 major pattern groups.

(a)

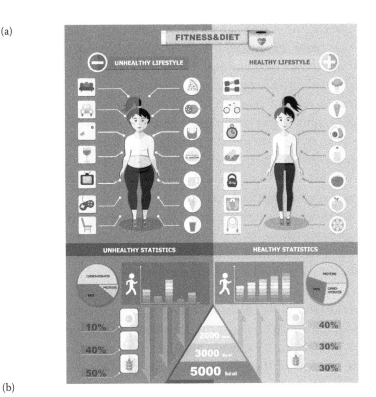

(b)

FIGURE 5.3  Examples for comparison (top, [17]) and concretize (bottom, [18]).

Data-driven storytelling is often about argumentation so it is important to be aware that there are different ways to present, support, reinforce, contradict, or discuss a given message. Patterns for argumentation serve the intent of persuading and convincing audiences.

The classical devices of rhetoric involve *logos* (reason, word), *ethos* (character, ideal), and *pathos* (experience, emotion). Though we believe the ultimate goal of data-driven storytelling is to communicate truth (most closely to logos), there are traces of both pathos (see the "Patterns for Empathy and Emotion" section), and ethos in every story, which help connect the narrator with the audience.

One very common pattern of argumentation is **compare** (Figure 5.3). The narrator presents two or more data sets, and draws respective conclusions from them. Data sets can refer to categories (e.g., healthy, unhealthy), elements in the data (e.g., food types), or temporal evolution. Visually, the comparison can be made through the juxtaposition of graphics (through side by side presentation), or through changes of a single graphic over time (e.g., animation). Comparison allows the narrator to make the point about equality of both data sets, to explicitly highlight differences and similarities, or to give reasons for their difference. If the graphic is rich enough, the audience can also be given the opportunity to explore differences and similarities for themselves.

**Concretize** is another pattern that can help build an argument by illustrating abstract concepts with concrete objects. Concretization usually implies that each data point is represented by an individual visual object (e.g., a point or shape), making them less abstract than aggregated statistics. Perhaps the most common example for concretization is the ISOTYPE language [18,19]. ISOTYPE links a specific pictogram (human, animal, cars, Figure 5.3(b)) to quantities (people fought in war, animals eaten per year, cars produced, water consumed [20]). Concretization has recently been used in interactive visualizations [21–23], and its effects studied in [24].

**Repetition** can also help argumentation, as it can increase a message's importance and memorability, and can help tie together different arguments about a given data set. Repetition can be employed as a means to search for an answer in the data. For example, the *Bloomberg* piece "What's Really Warming the World?" [25] shows different possible culprits behind global warming as a line chart depicting the evolution of each measure over time, such as volcanic or solar activity, deforestation, or ozone pollution.

While none of these measures correlates with the increase in global temperature, the last example, greenhouse gases, reveals a high correlation.

## Patterns for Flow

Patterns for flow are devices that help structure the sequencing of messages and arguments. They are essential to every story, as they set the order, rhythm, and pace, and help buildup the climax and resolution.

One common pattern for flow is **reveal**, in which elements of the data are progressively revealed, eventually leading to the whole picture and the final argument. Famous implementations of this pattern include the data videos on the Democracy.info site [26], and on "The Fallen of World War II" [27]. In this last piece, the narrator guides the audience through various sections of the data, starting with the fallen soldiers from the United States, then Poland, then France, etc., while sticking to the temporal order of events and battles. The narrative reaches its climax when the huge number of fallen Soviet soldiers is revealed; as more and more dead soldiers are stacked on top of each other—in the form of pictograms—the camera moves up in a seemingly endless vertical panning. The number of deaths seems infinite. By doing this, the narrator provokes strong emotional responses in viewers. It should be noted, though, that this particular example uses **reveal** in conjunction with **concretize** to achieve this highly effective outcome.

Reveal has also been used in *Bloomberg's* "What is Really Warming the World?" as each line in the line chart is revealed one after the other. Moreover, *gradual* reveal and **repetition** are closely related, and both can be seen to create rhythm and flow. Yet, repetition can happen without revealing new insights or data, and gradual reveal can happen without being repetitive. An example is the rhythm of slowdowns and breaks in Hans Rosling's narration of the world development data over 200 years [28]. Rossling slows down the pace of his animations as he focuses on dramatic changes during specific periods (here, World War II), and completely stops the animation to be able to explain the data in detail.

Narrative time in general can be a powerful means. In the example of "The Fallen of World War II" [27], presentation of data elements happens at a constant pace which supports the impression of the number of data elements. The counting of data elements in this particular example also provides a **slowing down** effect. The narration is slowed down and focused on a single aspect of the data, while leaving the user alone with the data.

In classical storytelling, a slowdown of the narrative pace can also happen through brakes.

The opposite to slowdowns and breaks is **speeding up** the narration or the presentation of data, which can be used for a similar effect. The example "U.S. Gun Deaths" [29] presents the observer with over 10,000 data points (individuals) and speeds up their reveal after an initial slow animation that explains how to read the visualization. The visualization uses the metaphor of bullets and ballistic arcs to represent people's expected lifespans, and the age at which they were shot. Due to both the metaphor and the effect of a sped-up, gun-fire-like presentation, the result is a strong emotional ensemble that creates an overwhelming effect based on data.

## Patterns for Framing the Narrative

Framing builds the way facts and events in a story are perceived and understood through narration. It feeds on the audience's expectations, but it can also play with those expectations and go against them to create surprise. Patterns for framing can be used to integrate the audience to the story, or conversely to keep it at a distance. They range from elements of the story itself, that help the audience relate to the content, and interaction techniques that give the audience control over the way the story unfolds, to purely narrative devices that challenge the audience and the way it is addressed.

Creating a **familiar setting** can help audiences relate to the content; it sets an entry point in the story that the audience can identify with. For example, the *OECD's Regional Well-Being* website [30] allows viewers to first inspect well-being in their region of the world, before moving on to explore and compare the quality of life in other regions. Similarly, inviting viewers to **make-a-guess** about the data or the outcome of a series of events presented in the narrative can enable them to take control over the way the story unfolds. This usually relies on an intelligent and original use of interaction. In the "You Draw it Yourself" example from *The New York Times* [31], readers are explicitly asked to draw the trend they expect to see in the data. This can engage them in causal reasoning about the issue addressed, which can stimulate their curiosity, and possibly lead to a game-like experience.

Conversely, **defamiliarization** challenges audiences' expectations, by presenting something familiar in an unexpected way. A typical example is to show a Mercator projected map (e.g., a Google Map) upside down. This challenges reading habits, and encourages exploring the map in new ways.

This pattern can be used to highlight and question implicit assumptions, and to force thinking differently about well-known facts. **Convention breaking** can also be used to similar effect by first establishing a graphical convention through, e.g., color, scales, or visual rhythm in the narrative, and then disrupting it at a specific point. This can highlight unordinary patterns or subsets in the data presented, and can create a surprise effect in the audience. Finally, introducing **silent data**, i.e., deliberately hidden data, can help create rhythm, like pauses in a piece music. The audience can use these pauses to try to infer the missing data, or to stitch different pieces of information back together.

From a purely visual perspective, a meaningful use of space and **physical metaphors** can help reinforce positive or negative feelings. Frequently used metaphors are: up = good, down = bad, left = retreat, and right = progress [32]. These take advantage of conventions to ease understanding, and can convey cultural or embodied messages.

## Patterns for Empathy and Emotion

Emotion and empathy are critical in storytelling. In fiction, emotional responses help create what the author Guy Gavriel Kay [33] terms *imaginative empathy*: resituating and reorienting the reader's perspective while enhancing our ability to understand and share the feelings and experiences important in the story. Empathy engages us with the story content: we pay attention and we perceive and reflect on the message. In fictional stories, empathy and emotions are evoked through realistic characters and situations; technology has provided us with immersive experiences through sound and stereovision. Technology and multimedia can enrich the audience's connection to the story through pure sensory experience, for focus, for compassion, for joy, and for excitement. However, data-driven stories—in the ideal case—do not rely on fictional content, or on ludic engagement, but are meant for education and to convey insight. What, then, are the specific applications and purposes of emotions and empathy in data-driven storytelling? How can they be used to enhance understanding, provoke a call-for-action, direct the audience to reflect on the data, or, more generally, enrich the persuasive message of the data story?

Some of the previously mentioned patterns, e.g., **gradual reveal** or **slowing down**, can contribute to emotional responses like surprise or the feeling of being lost. Managing flow can affect a sense of urgency, intensity, optimism, or seriousness. For example, the **speedup** in the onset of trails in the "U.S. Gun Deaths" piece [29] (Figure 5.4) heightens the sense

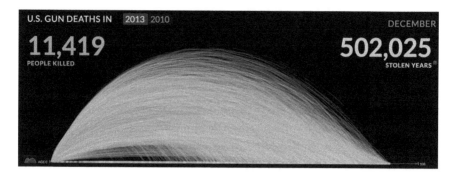

FIGURE 5.4 Example for **speed-up**; the way the individual traces are displayed, giving the impression of arms firing [29].

of urgency as the visualization progresses. Similarly, the measured pace of transitions in "What's Actually Warming the World?" [25] suggests a deliberate and methodical tone, invoking a sense of gravity and seriousness. These emotions can help engage the audience with the story's communicative intent (Figure 5.5).

FIGURE 5.5 Breaking-the-fourth-wall: *The Three Pigs* by David Wiesner [35].

**Concretize**, used in combination with other patterns [34], is another example that can strongly evoke emotions, grounding an empathic response within that emotional context. Consider the example of the human icons in *The New York Times* visualization of casualties in Iraq [36]. Using a sober black avatar of a mother and child to represent civilians moves the perception from a numerical assessment to the stark comparison of familial comfort, love, and security with the horror of war. Another intense example for a story provoking emotion is *Netwars* [37]. This example uses a pattern known from comics and movies; in **directly addressing the audience** and talking about his personal computer and his data, the narrator is **breaking-the-fourth-wall**. Originating from theater—the 4th wall is the wall between the stage and the audience, while the other 3 walls frame the scene (Figure 5.5)—**breaking-the-fourth-wall** can be a strong device for immersion into the story.

Another common pattern used to increase empathic ties between audiences and the story is **humans-behind-the-dots**. This pattern is similar to concretize, and is often used in journalism, where a concrete person or setup is introduced as an example for the matter of the story. Humans-behind-the-dots consists in presenting individual stories through detail-on-demand in a visualization. For example, the CNN story on U.S. casualties in the wars in Afghanistan and Iraq [38] shows casualties as dots on two maps. Clicking on the dots reveals the identities of each person, including their name, photo, hometown, as well as a description of what exactly happened to them. While pictograms or shapes can be used to concretize numbers, these details can highlight the uniqueness of each individual character and their life's story.

Related to humans-behind-the-dots is a pattern we call **familiarize**. This pattern helps create a familiar setting for audiences so that they can relate pieces of the data to what they already know (e.g., the situations in which they live). In publishing and explaining data on human development (education, health, income, etc.) for different geographic regions, the *OECD's Regional Well-Being* [30] website starts by asking viewers for their current location; it then presents the development data for that location, and shows regions with similar development profiles (Figure 5.6).

## Patterns for Engagement

Engagement can be seen as the feeling of being part of the story, of being connected to it, and being in control over the interactions with the story's content.

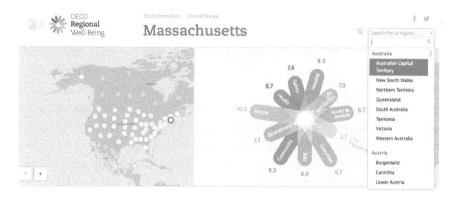

FIGURE 5.6    Example for the **familiarize** pattern: OECD Better Life Index [30].

One way to provoke (passive) engagement can be through emotion. Another way is to start with a **rhetorical question** (Figure 5.7). By directly addressing the audience (see **breaking-the-fourth-wall**), a rhetorical question can connect the narrator with the audience, and can trigger its will to answer that question, even if only for a moment. In this sense, a rhetorical question can set the context for a subsequent exploration of the reasons behind the answer, which the narration can then reveal and explain. Examples for rhetorical questions in data-driven stories are widespread, e.g., *What If I told you: you eat 3496 liters of water?* in Angela Morelli's "Virtual Water" [20], or "Can You Live on the Minimum Wage?" [39].

Sometimes, a rhetorical question can also imply a **call-to-action**. After having learned about how much water the production of certain food requires, Morelli [20] finishes with the words "Can we help?" and a list of action items for reducing personal water consumption, e.g., by regarding which food to eat, and how often. While many data-driven stories may intend for a call-to-action on different scales (e.g. "What's Really Warming the World" [25], or to reflect on war casualties [38]), these are not always explicit, nor are they necessarily coupled with concrete behavioral change takeaway (as is the case in Morelli's example).

Besides passive engagement in the form or emotion, empathy, or questioning the audience's own behavior, *active* engagement happens through audiences' action; it can shape the way the story is told, and which data is revealed (first).

Active engagement usually requires interactive features in a story presentation, if not a live presentation. For example, other than rhetorical

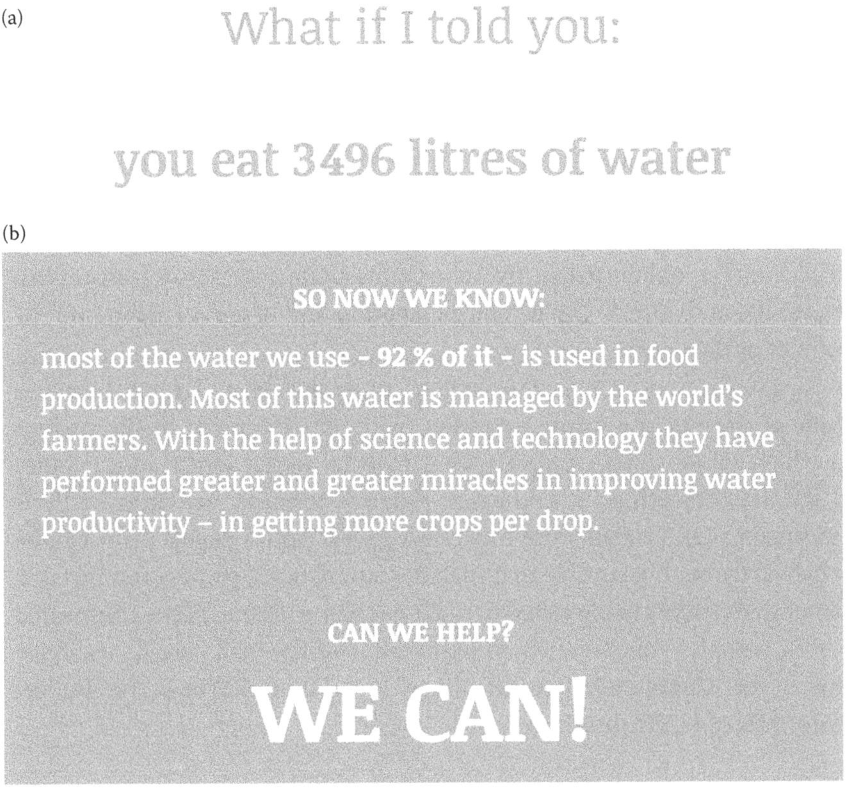

(a)

What if I told you:

you eat 3496 litres of water

(b)

SO NOW WE KNOW:

most of the water we use – **92 % of it** – is used in food production. Most of this water is managed by the world's farmers. With the help of science and technology they have performed greater and greater miracles in improving water productivity – in getting more crops per drop.

CAN WE HELP?

WE CAN!

FIGURE 5.7 Examples for patterns: **Rhetorical question** (top) and **call-to-action** (bottom). Both pictures from [20].

questions, actual questions can ask for an audience's opinion [37], which may potentially alter the narration of the story. The inclusion of personal data, such as location [30,40] or body weight [41], can also help interactively tailor the narration for the audience.

Alternatively to asking for actual data, the audience can be asked to **make-a-guess** about the provided information (e.g., about values or distributions). One example for this pattern is *The New York Times'* story on education [31]; the viewer is asked to draw a line inside a graph, representing the distribution between household income and percentage of children with a college degree. Make-a-guess questions can stimulate the audiences' curiosity to know the answer, and can engage them in causal reasoning about the phenomenon, while providing a motivation for following the narrative. On the one hand, audiences are asked to question their own perception of reality, and on the other hand, they are asked to question why this perception may have been wrong—in the case of a mismatch between their perception of reality and the actual data. Eventually, the narrative can also reveal what other audiences have guessed (a device used in the example of the college degrees, as well).

Make-a-guess questions can be posed in both remote and colocated presentation setups. In a colocated setup, the narration can be interrupted to address the audience directly. In a remote setup, interactive features can enable the audience to input their perception and understanding into the content of the narrative.

A last form of active engagement is **exploration**. Exploration can stand at the end of a linear narration (without user input) [8,42], such as in which viewers can interact with a map to display different attributes of the data depicting the countries' progress towards ratifying the Paris agreement [43]. Probably the highest degree of active engagement can be achieved through narrative that includes simulations, games, and puzzles; rather than simply being asked to make a guess, the audience is confronted with a complex interface where parameters of the data can be set freely (e.g., in the "Climate Change Calculator" [44] (Figure 5.8) or in the "Budget Hero" piece [45] (Figure 5.9)).

## USE CASES

We now cover three examples of data-driven narratives to illustrate how different narrative patterns can be used in combination.

### Case 1: U.S. Debt Visualized

Our first example talks about the scale of large numbers and how to communicate huge amounts of money. Demonocracy.info created a story to visualize the total debt of the United States [26]. This piece combines several narrative design patterns woven into a static website that requires scrolling to follow the story. Each step consists of an image and explanatory text (Figure 5.9). The same story with the same narration also exists as an online video [26].

The example starts with an image of a $100 bill, **concretizing** a certain amount of money into a tangible object of a certain size that everyone is familiar with. The next stage in the story shows a stack of 100 $100 notes bundled into a package that represents $10,000. The explanatory text states that this is the amount needed to buy a used car, or the equivalent of what the average human on earth earns in a year. The amount of $10,000, or a

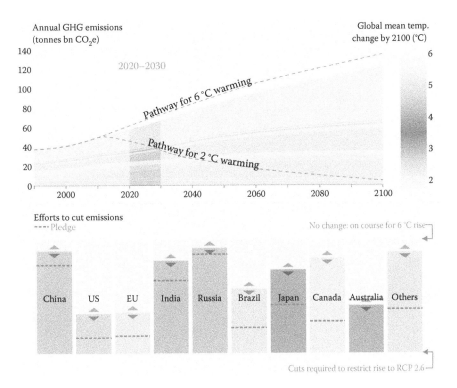

FIGURE 5.8 Examples for exploration and simulation with the Climate Calculator [44].

car, or 1 year worth of work, is concretized into a small package of paper easily fitting the pocket of a jacket. The next pictures scale the amount to $100 million, using the analogy of money stocked on a common pallet that usually is used to carry and lift goods. After another few pictures and increases in scale, the money shown grows to $1 trillion, which largely exceeds the surfaces of an American football field and of the White House in pallets of $1 million each. Eventually, the number of $122 trillion amounts into multiple skyscrapers, exceeding the height of the most important buildings in New York.

Thus, while **concretize** is predominant in this piece, it is not the only patterns used in this story. *Familiarization* is used through the different scale comparisons, juxtaposing and combining the number of bills and their volume with common spatial structures or material objects (e.g., pallets, a football field, and skyscrapers) and other immaterial values (e.g., that of a car, or of 1 years worth of work). This helps relate the numbers to things we can easily imagine, and gives a reference to other amounts of money.

(a)

(b)

(c)

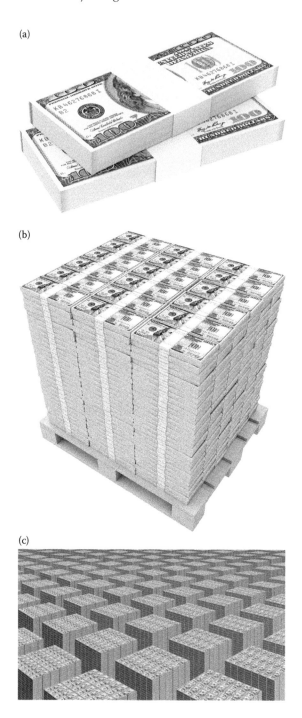

FIGURE 5.9   Case 1: U.S. debt ceiling visualized as different stacks of $100 bills [26].

The narrative also uses **repetition** and **gradual real**, as it is divided into individual stages that share the same structure, and present an increasing amount of money. This introduces a sense of flow to the narration.

## Case 2: Can You Live on the Minimum Wage?

This example is an interactive piece from *The New York Times* that reflects upon living on the minimum wage [39]. The piece briefly begins with some context, mentioning the number of workers currently living on the lowest legal pay in the United States. It then moves on to invite readers to see whether, given their current lifestyle, they could make ends meet with such an income. The hypothetical yearly income is **concretized**, using a unit visualization in which each unit represents $1. Units are displayed as small green squares piled from bottom to top. To the left of this visual representation are a number of input fields, in which readers are asked to indicate how much they spend on basic necessities like rent, transportation, food, etc. Every time a new amount is entered, the units in the visualization are removed to show how much money is left. If no money is left, a negative amount of units appears—shown with red squares piled from top to bottom—indicating that the reader would end up in debt by the end of the year.

This example also combines several narrative patterns. The first two are in the title of the piece itself: it **directly addresses the audience** by using a **rhetorical question**. The title asks whether "you" can live on the minimum wage, encouraging the reader to enter his personal expenses. The piece then serves to determine to what extent readers could not live with a minimum wage income. In addition, as the piece adapts to readers' input, it helps them to **familiarize** themselves with the amount of money that the minimum wage represents, by enabling them to compare it to the amount they spend on everyday necessities. This helps readers immediately identify with the story and the issue it addresses, i.e., the difficulties that 4.8 million Americans who live on the minimum wage have. Finally, readers can modify their entered values in order to *explore* through *repetition*.

## Case 3: What's Really Warming the World?

This last example is another interactive Web piece created by the *Bloomberg* graphics team [25] (Figure 5.8). It sequentially interrogates possible explanations for global warning in an attempt to debunk skeptics' arguments against manmade climate change. The piece starts with a **rhetorical question**: "What's Really Warming the World?," and confronts it with a time series line graph that shows the observed general increase in land-ocean temperatures between 1880 and 2014. Scrolling down the page reveals different trends for possible explanations and relates them to the general increase in temperature. The question also updates to, e.g., "Is it the Earth's Orbit?," "Is it the Sun?," or "Is it Volcanoes?," implying the reader should **make a guess**, based on the correlation he or she might see in the graphic. Of course, there are none, or at best, they are very weak. After going through several possible *natural* explanations, the questions and data switch to *man-made* explanations. This use of **repetition** is very powerful, as it pushes readers to try and **find answers for themselves.** While deforestation, ozone pollution, and aerosol pollution still show no real correlation, greenhouse gases show a very strong correlation. The question then changes to a direct statement, **breaking-the-fourth-wall**, telling the reader (and skeptics): "No, it really is greenhouse gases." The piece finishes by aggregating data for all possible man-made explanations, and the correlation becomes even stronger.

## DISCUSSION

We have described a set 18 prominent narrative patterns in the "Narrative Patterns" section, which we have categorized in 5 groups. These groups can help inform us as to when to use a certain pattern, i.e., what might be a potential purpose for the pattern. The section entitled "Use Cases" then provided a closer look at how different narrative patterns can be woven into a single narrative. While we believe this is an interesting start, we are also convinced that several open questions remain, e.g., how can we employ our collection to create our own narratives? or, how are patterns affected by the presentation medium or the audience?

### Using Narrative Patterns

We think narrative patterns can be used in both a generative and analytical fashion by a wide variety of storytellers: journalists, teachers, students, designers, visualization researchers, scientists, etc.

**Generative use:** Our main intention is to provide a pool of possibilities and ideas that help authors create data stories. Like other examples of design cards [10,11], narrative patterns can provide a starting point and initial ideas to assist and augment the creative process. In a typical case, cards can be randomly selected, and their respective narrative patterns can be used to shape a tentative narrative for a given story. Other cards can then be selected to produce alternative narratives, creating diversity and potentially triggering discussion.

**Analytical use:** Narrative patterns can also help to analyze narratives of existing data-driven stories by encouraging researchers and critics to think about the specific intents behind the use of different patterns. The patterns could typically be used in education to teach data-driven storytelling by analyzing which patterns go well together in a specific story?, which patterns exclude each other?, or, which are the favorites?

Eventually, the process of creating new narratives may lead to novel patterns or novel variants of existing patterns. We envision our corpus of narrative design patterns will grow organically, enriching both the creation of new narrative experiences as well as our understanding of the use of narrative techniques in data-driven storytelling.

## Storytelling Techniques

Narrative patterns are not storytelling "techniques" themselves. This means, our patterns are not technical blue prints or code that can be downloaded and applied in a straightforward manner such as User Interface Patterns [46], Visualization Design Patterns [47], or Software Design Patterns [48]. Narrative patterns do not imply any visual (or other sensory) form, or even interaction techniques; the eventual visual, auditory, and interaction design remains to be created for each story individually and benefits from (and requires) the creativity and expertise of the author.

## Presentation Medium

Since narrative patterns are independent from the eventual presentation, they are also generally independent from the presentation medium— Web, oral presentation, print, video, and so forth. Some patterns may not apply to specific media. For example, patterns for engagement might require a back channel between the audience and the storyteller. Emotion can potentially be transported better through certain media (e.g., movies or sound, as in *Out of CTRL* [37]). We have consistently talked about the storyteller and the audience. Of course, depending on the presentation medium, the storyteller can be colocated with the audience; or having authored the story content, the audience may consume it remotely and asynchronously. We think there are vast possibilities for future research here.

## Different Notions of Time

Talking about the presentation medium and the potential colocation of storyteller and audience, we necessarily have to talk about the notion of time in (data-driven) storytelling. We refer to three notions of time: *authoring-time*, *presentation-time*, and *data-time*.

**Authoring-time:** Authoring-time refers to the time when the story and its narration is created by the author. In some cases, story and narration are detached in time, i.e., the story is usually written before the narration and can be written by another person. In any case, we consider both as authoring-time. During authoring-time, patterns are "selected" and/or "implemented" into the final narration.

**Presentation-time:** Presentation-time specifies the time at which a presenter presents the story to an audience, or the audience consumes the story, depending on the presentation medium. Generally, patterns are not changed during presentation-time. However, there may be cases where the audience may choose its own narration through the story, through interaction, and through exploration. Also, patterns such as repetition can be implied in an exploratory narration; for example "Can You Live on the Minimum Wage?" [39], as users play with different budgets, expenses, and geographical regions.

**Data-time:** Finally, the third notion of time is the time indicated in the data itself. For the sake of communication, presentation-time can be related to data-time. One such example is the counting of soldiers in "The Fallen of World War II" [27]; the time of people being killed in the war is implicitly mapped to the time as the video continues while more and more data points (soldiers) are added to the screen. A similar example relating presentation-time and data-time is presented by Hans Rosling talking about the development of nations [28]; Rosling slows down, speeds up, and stops the pace of the data as it is animated through the changing years. Interestingly, his or her voice reflects the same pace as the data changes, slow when there are small changes and speeding up as the data begins to change more rapidly.

Besides animation, visual techniques for presenting time are well-studied in the literature; Aigner et al. [49] survey general visualization for temporal data, Brehmer et al. [50] overview timeline visualizations, Beck et al. [51] compile an overview over visualizations for time-chancing networks, and Bach et al. [52] describe spatiotemporal data such as geographical data, videos, networks, multidimensional data, and others. Visualizations and systems specific for presenting temporal data are only in the beginning of gaining greater attention [50,53,54].

## Audience and General Intention

Similar to the presentation medium, the target audience can vary—children, students, experts, car owners, and consumers. Certain patterns may work for a specific audience (e.g., *rhetorical questions*, *make-a-guess*, and *exploration*), while not for other types of audiences. Audiences differ in homogeneity, in background knowledge, in their willingness to interact, in cultural behaviors, and in individual views on the world. Depending on the audience, a storyteller may have different motivations for communication (informing, educating, confirming, questioning, moving to act, etc.) and, hence, change the narrative patterns that they choose. Sometimes an author may have the luxury of preparing multiple versions for different audiences (or media) while at other times, they may need to fall back on a lowest common denominator. See Chapter 9 entitled "Communicating Data to an Audience" for more details.

## CONCLUSIONS

This chapter introduced narrative patterns for data-driven storytelling. We described and illustrated a core collection of 18 patterns with examples, discussed relations between patterns, and gave scenarios on their usage.

We are only at the very beginning of understanding how data-driven storytelling works and how narrative patterns can benefit the communication of a message. Beyond creating data-driven stories, we hope that narrative patterns will help authors think more broadly about data-driven storytelling. For example, the discussion of what can be called data-driven storytelling is far from settled (hence, the motivation for this book), and the transition from explanations based on illustrations and graphics towards data-driven argumentations is gradually sinking in. Also, we do not think (all) narrative patterns are limited to data-driven storytelling. Rather, we got inspiration from common storytelling mediums and looked for common patterns across the domains, i.e., where the rather novel medium of data-driven stories is related to the classical mediums. We also do not see narrative patterns bound to a specific technical medium such as videos, static or interactive graphics, interactive presentations, games, news articles, etc. We hope that the narrative patterns presented here can contribute to the discussion on what data-driven storytelling is, what its mediums are, and what are good or bad practices.

Many questions remain to be answered. What are the specific rules a specific narrative pattern should adopt or implement? Which patterns go well with each other and which exclude each other? Though we do not aim to restrict and over-quantify narrative patterns, such questions can serve to establish a common ground to create effective stories.

Other questions remain on which patterns are suitable for which audience and how particular patterns apply to specific media; how does the presentation medium change the implementation of the narrative pattern? This is closely related to the specific techniques that are used in the implementation of a narrative pattern, and how those are related to the presentation medium, e.g., animation, interactive highlighting, multiple choice questions, interactive simulations, etc. So far, the links between narrative pattern and presentation technique remain unclear. We are just beginning to think about the respective implications and purposes of each pattern. How persuasive is a narrative pattern? Can we measure effectiveness? Is a pattern more emotional or more rational? Does it aim for objectiveness or for provoking an emotion? Potentially, there are alternative groupings than the ones proposed in the "Narrative Patterns" section. Eventually, there are many open questions related to the higher-level structure of data-stories, including rhythm and pace. Are there specific patterns for story-structures, similar to the structures of classic Greek drama, comic books and graphic novels, documentaries, or scientific articles?

While there are an infinite number of possibilities to combine, adapt, extend, and implement patterns, we think our initial collection is a good starting point to discuss all these questions and to start better understanding data-driven storytelling. We hope for a growing collection of patterns, as well as for a creative discussion in the near future.

## REFERENCES

1. G. Genette, *Narrative Discourse: An Essay in Method*. Cornell University Press, Ithaca, 1983.
2. J. S. Bruner, *Actual Minds, Possible Worlds*. Harvard University Press, Cambridge, MA, 1985.
3. G. Branston and R. Stafford, *The Media Student's Book*. Psychology Press, London, 2003.
4. N. W. Kim, B. Bach, H. Im, S. Schriber, M. Gross, and H. Pfister, "Visualizing nonlinear narratives with story curves," *IEEE Transactions on Visualization and Computer Graphics (TVCG)*, vol. 24, no. 1, pp. 595–604, 2018.
5. S. McCloud, *Understanding Comics: The Invisible Art*. Kitchen Sink Press, Northampton, MA, 1993.
6. B. Alexander, *The New Digital Storytelling: Creating Narratives with New Media: Creating Narratives with New Media*. ABC-CLIO, LLC, Santa Barbara, CA, 2011.

7. N. Gershon and W. Page, "What storytelling can do for information visualization," *Communications of the ACM*, vol. 44, no. 8, pp. 31–37, 2001.

8. E. Segel and J. Heer, "Narrative visualization: Telling stories with data," *IEEE Transactions on Visualization and Computer Graphics*, vol. 16, no. 6, pp. 1139–1148, 2010.

9. K.-L. Ma, I. Liao, J. Frazier, H. Hauser, and H.-N. Kostis, "Scientific storytelling using visualization," *IEEE Computer Graphics and Applications*, vol. 32, no. 1, pp. 12–19, 2012.

10. "Ideo cards." online: https://www.ideo.com/work/method-cards. [last visited: Nov. 29, 2016].

11. "Oblique strategies." online: https://en.wikipedia.org/wiki/Oblique_Strategies. [last visited: Nov. 29, 2016].

12. "Design with intent." online: http://designwithintent.co.uk/docs/designwithintent_cards_1.0_draft_rev_sm.pdf. [last visited: Nov. 29, 2016].

13. "Creativity cards." online: http://ja-ye.atom2.cz/form/download.ashx? FileId=52. [last visited: Nov. 29, 2016].

14. "Service design tools." online: http://www.servicedesigntools.org/frontpage? page=3. [last visited: Nov. 29, 2016].

15. S. He and E. Adar, "Vizitcards: A card-based toolkit for infovis design education," *IEEE Transactions on Visualization & Computer Graphics*, no. 1, pp. 561–570, 2017.

16. B. Bach, Z. Wang, M. Farinella, D. Murray-Rust, N. H. Riche, and W. A. Redmond, "Design patterns for data comics," in *Proceedings of ACM Conference for Human Factors* in *Computing Systems (CHI)*, 2018.

17. "Education vs. incarceration." online: http://visual.ly/education-vs-incarceration. [last visited: Nov. 29, 2016].

18. O. Neurath, *International Picture Language; the First Rules of Isotype: With Isotype Pictures*. K. Paul, Trench, Trubner & Company, London, 1936.

19. "Gerd Arntz Web Archive." online: http://gerdarntz.org/isotype. [last visited: Nov. 29, 2016].

20. A. Morelli and InfoDesignLab.com, "The water we eat." online: www. thewaterweeat.com. [last visited: Nov. 29, 2016].

21. F. Chevalier, R. Vuillemot, and G. Gali, "Using concrete scales: A practical framework for effective visual depiction of complex measures," *IEEE Transactions on Visualization and Computer Graphics*, vol. 19, no. 12, pp. 2426–2435, 2013.

22. "Sand dance." online: https://www.sanddance.ms. [last visited: Nov. 29, 2016].

23. "The globe." online: http://globe.cid.harvard.edu/. [last visited: Nov. 29, 2016].

24. S. Haroz, R. Kosara, and S. L. Franconeri, "Isotype visualization: Working memory, performance, and engagement with pictographs," in *Proceedings of the 33rd Annual ACM Conference on Human Factors in Computing Systems*, CHI '15, pp. 1191–1200, ACM, New York, 2015.

25. "What's really warming the world?" online: http://www. bloomberg.com/ graphics/2015-whats-warming-the-world/. [last visited: Nov. 29, 2016].

26. "US debt ceiling visualized in $100 bills." online: http://democracy.info/infographics/usa/us_debt/us_debt.html. [last visited: Nov. 29, 2016].

27. "The fallen of World War II." online: https://vimeo.com/128373915. [last visited: Nov. 29, 2016].

28. "Hans Rosling's 200 countries, 200 years, 4 minutes—The joy of stats—BBC Four." online: https://www.youtube.com/watch?v=jbkSRLYSojo. [last visited: Nov. 29, 2016].

29. "U.S. gun deaths in 2013." online: http://guns. periscopic.com. [last visited: Nov. 29, 2016].

30. OECD, OECD countries, United States, Massachusetts, from OECD Regional Well Being. online: https://www.oecdbetterlifeindex.org. [last visited: Nov. 29, 2016].

31. "You draw it: How family income predicts children's college chances." online: http://www.nytimes.com/interactive/2015/05/28/upshot/you-draw-it-how-family-income-affects-childrens-college-chances.html. [last visited: Nov. 29, 2016].

32. "The globe." online: https://www.theguardian.com/world/ng-interactive/2015/aug/26/china-economic-slowdown-world-imports. [last visited: Nov. 29, 2016].

33. "A conversation with Guy Gavriel Kay." online: http://www.fantasy-magazine.com/non-fiction/articles/a-conversation-with-guy-gavriel-kay. [last visited: Nov. 29, 2016].

34. J. Boy, F. Detienne, and J.-D. Fekete, "Storytelling in information visualizations: Does it engage users to explore data?" *Proceedings of the 33rd Annual ACM Conference on Human Factors in Computing Systems*, CHI '15, Seoul, Republic of Korea, New York: ACM, pp. 1449–1458.

35. D. Wiesner, *The Three Pigs*. Random House, New York, 2014.

36. A. L. de Albuquerque and A. Cheng, "31 days in Iraq." online: http://www.formfollowsbehavior.com/2007/09/17/artistic-data-based-visualization/31 days_iraq_20061/. [last visited: Nov. 29, 2016].

37. "Out of CTRL: Die Web-Doku." online: http://future.arte.tv/de/netwars-krieg-im-netz-1/out-ctrl-die-web-doku. [last visited: Nov. 29, 2016].

38. "Casualties." online: http://edition.cnn.com/SPECIALS/war.casualties/index.html. [last visited: Nov. 29, 2016].

39. "Can you live on the minimum wage?" online: http://www.nytimes.com/interactive/2014/02/09/opinion/ minimum-wage.html. [last visited: Nov. 29, 2016].

40. "The division game." online: http://collapse-thedivisiongame.ubi.com. [last visited: Nov. 29, 2016].

41. "Your Olympic athlete body match." online: http://www.bbc.com/news/uk-19050139. [last visited: Nov. 29, 2016].

42. J. Boy, F. Detienne, and J.-D. Fekete, "Storytelling in information visualizations: Does it engage users to explore data?" in *Proceedings of the 33rd Annual ACM Conference on Human Factors in Computing Systems*, CHI '15, pp. 1449–1458, ACM, New York, NY, USA, 2015.

43. "When could the Paris Agreement take effect? Interactive map sheds light." online: http://www.wri.org/blog/2016/04/when-could-paris-agreement-take-effect-interactive-map-sheds-light. [last visited: Nov. 29, 2016].

44. "Climate change calculator." online: http://ig.ft.com/sites/climate-change-calculator. [last visited: Nov. 29, 2016].

45. "Budget hero." online: https://www.wilsoncenter.org/budget-hero. [last visited: Nov. 29, 2016].

46. "User interface patterns." online: http://ui-patterns.com/patterns. [last visited: Nov. 29, 2016].

47. "Visualization design patterns." online: http://www.infovis-wiki.net/index.php?title=Visualization_Design_Patterns. [last visited: Nov. 29, 2016].

48. E. Gamma, *Design Patterns: Elements of Reusable Object-Oriented Software*. Pearson Education India, London, 1995.

49. W. Aigner, S. Miksch, W. Müller, H. Schumann, and C. Tominski, "Visualizing time-oriented data—A systematic view," *Computers & Graphics*, vol. 31, no. 3, pp. 401–409, 2007.

50. M. Brehmer, B. Lee, B. Bach, N. H. Riche, and T. Munzner, "Timelines revisited: A design space and considerations for expressive storytelling," *IEEE Transactions on Visualization and Computer Graphics*, 2016.

51. F. Beck, M. Burch, S. Diehl, and D. Weiskopf, "The state of the art in visualizing dynamic graphs," *EuroVis STAR*, 2014.

52. B. Bach, P. Dragicevic, D. Archambault, C. Hurter, and S. Carpendale, "A descriptive framework for temporal data visualizations based on generalized space-time cubes," in *Computer Graphics Forum*, vol. 36, no. 6, pp. 36–61, 2016.

53. B. Bach, N. Kerracher, K. W. Hall, S. Carpendale, J. Kennedy, and N. H. Riche, "Telling stories about dynamic networks with graph comics," in *Proceedings of the Conference on Human Factors in Information Systems (CHI)*, ACM, New York, 2016.

54. B. Bach, C. Shi, N. Heulot, T. Madhyastha, T. Grabowski, and P. Dragicevic, "Time curves: Folding time to visualize patterns of temporal evolution in data," *Transactions on Visualization and Computer Graphics*, vol. 22, no. 1, pp. 559–568, 2016.

# CHAPTER 6

# Watches to Augmented Reality

## Devices and Gadgets for Data-Driven Storytelling

Bongshin Lee

*Microsoft Research*

Tim Dwyer

*Monash University*

Dominikus Baur

*Independent Researcher*

Xaquín González Veira

*Independent Consultant*

## CONTENTS

## INTRODUCTION

In visual data-driven storytelling, hardware devices and media with which stories are delivered play an important role, making a significantly different impact on both presentation and consumption experiences. The different form factors offer different affordances for data storytelling, affecting their suitability to the different data storytelling settings (Lee et al., 2015; see Chapter 7, entitled "From Analysis to Communication Supporting the Lifecycle of a Story"). For example, wall displays are well-suited for synchronous colocated presentations (i.e., where the presenter and audience are together at the same time, in the same room), while wristwatches and virtual reality headsets work better for personal consumption of pre-authored data stories. The title of this chapter includes "Devices *and* Gadgets" to avoid confusion with the other kind of "storytelling devices" (i.e., narrative techniques; see Chapter 4, entitled "Data-Driven Storytelling Techniques: Analysis of a Curated Collection of Visual Stories").

In this chapter, we discuss a list of different device form factors as well as their affordances and characteristics for different storytelling settings. We consider a fairly wide range of form factors for data-driven storytelling, including not only the obvious electronic "devices," but also more diverse media such as tangible props (i.e., things that people can pick up and hold, gesticulate with, and so on). The latter are worth considering because they can give insights into how data storytelling might occur in futuristic mixed-reality digital environments that may be enabled by the current rapid progress in virtual reality (VR) and augmented reality (AR) display and interaction technologies. In addition, the devices we consider not only display content but also offer direct interaction in most cases. We also present a set of examples that use different devices for data-driven storytelling, reflecting on how the technology enables or otherwise influences storytelling. We conclude this chapter with a discussion of the evolution of display and interaction device technology, and how it might affect future data-driven storytelling capabilities.

Note that we do not consider the story creation (i.e., authoring) experience because it seems likely that, for the foreseeable future, creation of engaging visual data-driven stories will require powerful desktop platforms. We focus on the story presentation and consumption experiences, specifically considering the following two scenarios. First, a presenter gives a live (i.e., synchronous) presentation to the audience. We assume that a single person (at least one at a time) presents the story. Second, the audience consumes the data story without any live presenter (i.e., asynchronous). For both cases, the audience may be different sizes, from one person to a group of people.

## CHARACTERISTICS OF DIFFERENT DEVICES

In this section, we delineate a set of dimensions to better characterize the affordances of different devices. We propose six dimensions: live presentation capability, perception size, level of immersion, portability, level of isolation, and haptic feedback (Table 6.1).

By **live presentation capability**, we refer to the dynamics or relationships between the presenter and the audience for a live presentation, and we give an estimate for an audience size that can comfortably view the device simultaneously. A small display like a phone screen is "personal" but not "private" because it is primarily intended for one person, but can be shown to one or two close onlookers. A "shareable" device allows for a presenter and an active audience to interact with the device. That is, we make the distinction between "passive" and "active" audiences, where a passive audience does not directly interact with the device while an active audience is able to interact with the device themselves to, for example, change the view or filter the data. Watches are private because they are inherently less shareable than other small-screen devices such as phones. They are typically physically attached to the owner's wrist and, hence, awkward for a presentation to a passive audience and nearly infeasible for active sharing.

VR and AR headsets provide a private view in that a headset can only be worn by one person and the typical story consumption scenario would be a single person experiencing the data visuals from the preparation by the author. However, it is worth noting that a virtual environment can be shared over a network by a large group of people, each wearing their own headset. Each person then has the freedom to move around the virtual data presentation, to change their viewpoint, and possibly to interact as an active audience.

Wall displays allow a presenter to give a live presentation to a group of people in a typical conference room setting. A pen- and/or touch-enabled wall display may also be large enough to allow a few people to share (i.e., synchronously interact with) the display. Tabletop displays are still more suitable to this latter-shared scenario in that people can stand or sit around the display and interact comfortably with the horizontal surface.

TABLE 6.1  Affordances and Characteristics of Different Devices

| Device | Live Presentation Capability | Perception Size | Level of Immersion | Portability | Level of Isolation | Haptic Feedback |
|---|---|---|---|---|---|---|
| Watch | Private (nearly infeasible for a live presentation) | Very small | Low | Yes | Low | Yes, low fidelity |
| Phone | Personal 1–2 passive audience | Small | Low | Yes | Medium | Yes, low fidelity |
| Tablet | Personal 1–3 passive audience | Medium | Low | Yes | Medium | Yes, low fidelity |
| Laptop PC | 1–5 passive audience (or large passive audience if projected) | Medium | Low/medium | No | Medium | No |
| Desktop PC | 1–5 passive audience (or large passive audience if projected) | Medium | Low/medium | No | High | No |

*(Continued)*

TABLE 6.1 (*Continued*) Affordances and Characteristics of Different Devices

| Device | Live Presentation Capability | Perception Size | Level of Immersion | Portability | Level of Isolation | Haptic Feedback |
|---|---|---|---|---|---|---|
| AR headset | Private display (shared view of the environment) Shareable virtual environment (each audience member has their own headset) | Unrestricted (current devices have limited field of view) | High | Yes | Low | No |
| Tabletop | Sharable 1–3 active audience 1–3 passive audience | Medium to large | Medium | No | Low | No |
| Wall display | Sharable Large passive audience 1–3 active audience | Medium to very large | Medium | No | Low | No |
| VR headset | Private display Shareable virtual environment | Unrestricted | High | No | High | No |
| Tangible | Sharable 1–3 active audience Large passive audience | Very small to very large | Medium/high | Depends | Low | Yes, high fidelity |

Traditional personal computer devices, whether a laptop or desktop, are designed for a single person. That is, traditional interaction through a mouse and keyboard is not really sharable. However, these devices do support a small, passive audience of onlookers.

**Perception size** refers to the size of the display relative to the viewer's field of view at the usual viewing distance. Perception size for most devices (including watches, phones, tablets, wall displays, and tabletops) is proportional to the device screen's physical size. On the other hand, AR and VR headsets achieve (relatively) unrestricted perception size.

Our use of the term "immersion" is inspired by the commonly accepted definition used in the field of VR and AR research, where it is usually used to describe how much you feel like you are a part of (or have *presence* in) the virtual environment (Bowman et al., 2005, p. 4). In addition, it is also inspired by more nuanced notions of immersion from the computer games nomenclature (Adams, 2004), for example, "narrative immersion" being somewhat synonymous with "engagement." Thus, we use the **level of immersion** dimension to describe how much of the audience's attention is required by the device, and how much it contributes to engaging the audience in the data-driven story being told.

By **portability**, we consider not only whether the device can be moved, but also whether a device can be used "on-the-go" and, for example, while the viewer is doing other things. Thus, a laptop can at least be moved but probably cannot be used to consume data stories while walking a dog. As the name implies, the **level of isolation** captures the degree from which the viewer is isolated from the real-world environment mainly due to the device. For example, VR headsets offer a high level of isolation while AR headsets deliver a low level of isolation. Lastly, **haptic feedback** can have a significant impact on the audience's consumption experience, stimulating their tactile sensation. Tangible devices inherently provide high fidelity haptic feedback.

Note that we do not include output resolution as a dimension because it is a moving target and is currently heading towards being so high that a human can no longer distinguish pixels (i.e., "retinal") across all devices. We also exclude accessibility because it is too complex and likely to require a different family of devices. Accessibility of data-driven stories to differently abled people is obviously very important, but the space of different needs that people have is huge and in general, this is a very understudied topic. There is very limited research on devices to support visualization accessibility. A rare example is the GraVVITAS system, which enables blind people to consume information graphics through sonification (the use of sound) and haptic feedback (Goncu and Marriott, 2011). More generally, there has been some study of tangible map creation for vision-impaired people (Ducasse et al., 2017). However, to the best of our knowledge, there is no holistic survey on the topic of accessibility for visualization or data-driven storytelling though there is a clear need for further research in this area.

## EXAMPLES OF PRACTICES BY DEVICES

Watches come with a very small screen, accommodating a private consumption. They are usually appropriate for single-sentence stories, providing linear narration (Figure 6.1). Word-scale visualizations, a more-general version of sparklines—small, word-sized data graphics that allow meta-information to be visually presented in-line with document text (Goffin et al., 2014)—can be embedded within the story for watches.

Phones are highly interactive and are mostly adequate for the consumption of many types of stories. Research shows that people currently spend much more time consuming digital media on the phone than larger screen devices (The Nielsen Company, 2014). We argue that it is more likely for people to encounter and consume data stories (such as those in Figures 6.1 and 6.2) in this small-screen form factor than other types of devices. It is also worth noting that, while we consider the live presentation capability

FIGURE 6.1  Due to a small screen size, stories on watches are usually presented with a single image or sentence (www.theguardian.com).

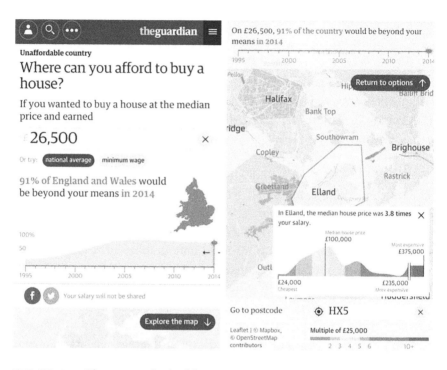

FIGURE 6.2 Phones can be highly interactive, providing a set of widgets for interaction (e.g., selection, filtering) and supporting a map navigation.

of the devices in Table 6.1 and in the following discussion, the default data-story consumption modality for these small-screen devices (including watches and tablets) is still asynchronous (i.e., off-line).

While not ideal for synchronous presentation (i.e., the perception size is limited), it is still possible to use phones with a small audience (one or two people). Even though linear narration is the default due to just enough screen real estate combined with simple touch interaction, it is possible to offer more sophisticated navigation (Figure 6.2). Thus, the audience may have more commitment and engagement to the story. As tablets are very similar to phones, they share characteristics with phones such as better accommodating a small audience (1–3 people). However, as they are bigger in size, synchronous presentation scenarios are more feasible on tablets than on phones. Furthermore, as digital pens are becoming more prevalent and effective, tablets can provide precise selection and hover, making them more useful for a live presentation than phones.

Desktops and laptops as well as wall displays provide a novel interaction and powerful navigation mechanism, supporting asymmetric storytelling in the sense that the audience is passive while the presenter "drives" (Figure 6.3). However, desktop and laptop PCs can be effective for only several audience members while wall displays can be comfortable for up to about a dozen audience members. These displays (especially laptops) can be projected to a big screen to support a large group of people (Figure 6.4).

With a (horizontal) tabletop, a presenter and the audience has a symmetric experience (i.e., peers) for colocated synchronous scenarios. However, the orientation of the display becomes an issue for symmetry for multiple people situated around the table. For example, the presenter may

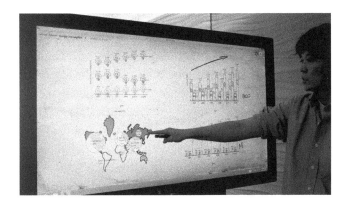

FIGURE 6.3 As demonstrated by SketchStory (Lee et al., 2013), wall displays supporting a novel pen and touch interaction allow a presenter to deliver a more-engaging presentation.

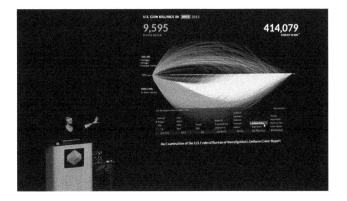

FIGURE 6.4 At OpenVis Conf 2013, Kim Rees presents a story about U.S. gun killings while discussing how her team created the story including the open data and open source tools they used to create it.

have to look at the visualization upside down in order for the audience on the other side of the table to see it right-side up.

In addition, devices that can track a person's entire body postures and gestures, such as the Microsoft Kinect, offer a presenter more engaging presentations on wall displays through the use of *proxemics* (Jakobsen et al., 2013). That is, when the system is aware of the presenter's precise position (and proximity) relative to the display, it can alter the data display as desired, for example, to show greater detail for a feature that the presenter is pointing to.

VR headsets afford full immersion, isolating the audience from the real-world environment including other people. It can be particularly engaging for a committed audience, and also possible to use depth and head and hand tracking to provide novel experiences. While the default VR experience is isolating and viewed privately, multiple VR headsets can be networked so that people can share and interact in a virtual scene to enable distributed or colocated synchronous storytelling scenarios (Figure 6.5).

AR devices offer exciting possibilities for future data storytelling scenarios. For example, Hans Rosling reveals the story of the world's past, present, and future development using "simulated" AR animation in a video presentation (Figure 6.6). This is simulated AR in that the augmented visuals have been carefully added to the video recording by animators, supporting asynchronous yet broad consumption. However, in the future, such an interesting and creative presentation might be possible in a live (i.e., synchronous) setting where the audience members each have an AR headset that shows the data visuals hanging in space in front of the

FIGURE 6.5  Telling a story about historical air traffic data in VR. The presenter (left person) explains to an audience member (right person) that aircraft trajectories form a "stack" of hold patterns near an airport. The audience sees the virtual scene from precisely the same viewpoint as the presenter and the presenter's hands are tracked so that his gestures are visible in the virtual scene (right). Courtesy Maxime Cordeil (Cordeil et al., 2016).

FIGURE 6.6 Hans Rosling tells the story of the world in 200 countries over 200 years using 120,000 numbers using (simulated) augmented reality animation.

presenter. Furthermore, the audience members may even be able to interact with the data visuals independently.

Creative presenters are increasingly using tangible objects in their presentations to foster audience engagement and improve audience understanding. For example, in his TED talk,* Hans Rosling explains why it is crucial to end poverty to control population growth using IKEA boxes. We note that artistic treatments (e.g., data sculpture) or free from data explorations (e.g., the navigable bar chart; Taher et al., 2017) are different from when physicalization with tangible objects is used for storytelling. Arguably, such data-inspired artworks are intended to invoke an emotion or even to mystify rather than simply engage.

## OPPORTUNITIES AND CHALLENGES

New and refined devices introduce opportunities and challenges to the visualization community both for exploration and presentation. Several visualization researchers have called for more efforts on developing visualization and interaction techniques that go beyond the traditional desktop environment (Lee et al., 2012; Isenberg et al., 2013; Roberts et al., 2014; Chandler et al., 2015). A wide range of screen sizes poses an important challenge: the need to design responsive visualization (Leclaire and Tabard, 2015) that can adapt to different form factors or simply to different orientations (i.e., portrait versus landscape).

We also need to address the well-known issues with 3D and virtual environments (Munzner, 2014, Chapter 6). For example, 3D environments

* https://www.ted.com/talks/hans_rosling_on_global_population_growth.

usually suffer from occlusion, the effect of one object blocking another object from view. Visually induced motion sickness can occur when the audience views drastic visual representations of self-motion or when detectable lags are present between head movements and presentations in a head-mounted display (HMD) (Hettinger and Riccio, 1992). However, rapid improvements in tracking and rendering technologies have reduced such latencies to the point where motion sickness is becoming much less of an issue (Carmack, 2013). Now that cardboard devices—costing just a few dollars—can turn a phone into a VR headset (e.g., Google Cardboard), VR is definitely a technology worth considering for delivering engaging data-stories to a mass audience. Observation of the way that people tell data-driven stories with physical props (like Hans Rosling's buckets and flip-flops*) suggests exciting ways that smart devices can be incorporated into tabletop (Dalsgaard, 2014), VR or AR experiences (Cordiel, 2017) to further engage through so-called "mixed-reality" experiences (Milgram and Kishino, 1994).

Willett et al. (2017) give a comprehensive taxonomy of the ways that data displays can be brought into the world using technologies like AR. The idea of delivering data-driven stories relevant to people's location and situation within their environment has enormous possibilities for communicating to museum visitors or tourists in a city. However, wearable AR HMDs have some further technological challenges to solve before being unobtrusive enough that people would want to wear them for extended periods in such settings. As well as needing to become smaller and lighter, the field of view needs to be increased, and a capability to allow the audience to focus on virtual objects at different depths must be added before truly realistic augmentation is achieved. The last issue (depth of field) is particularly challenging and possible solutions, while promising, have not yet left research labs (Huang et al., 2015).

We encourage researchers and practitioners to continue exploring ever-more creative means of telling data-driven stories leveraging these ever-evolving devices. As we have seen, there are more ways than ever before to communicate about data and there will be even more options as the technology continues to improve.

In this chapter, we limited our discussion to devices for data-story presentations and consumption rather than the hardware platforms upon which software tools (see Chapter 7, entitled "From Analysis to

---

* https://www.ted.com/talks/hans_rosling_on_global_population_growth.

Communication: Supporting the Lifecycle of a Story") for authoring sophisticated data stories are typically run. As stated in the introduction, we assumed that currently traditional personal computer devices are the main platform for such authoring. This is partly because a mouse and keyboard are still the preferred way to work with complex media-authoring software. But, another major factor is that smaller devices still have limited processing power in terms of onboard memory, graphics processing capability, and central processing unit (CPU) speed. Preparing data stories also usually requires exploration of very large data sets that would be beyond the capacity of small devices. In the future, it may be possible to close this division. The processing power of smaller devices may be less of a limiting factor, while the human factors of small screens and interaction capabilities will require more creative solutions.

## CONCLUSION

Over the decades we have seen the display devices available for computational data-driven storytelling both shrink to fit into a watch face and grow to cover walls while HMDs offering VR and AR experiences are now commodity devices. In parallel, we have witnessed the introduction of novel interaction techniques from touch to body tracking. These can be leveraged to engage the consumer through reactive data-driven storytelling, or by promoting the feeling of "being there" or being present. To provide better presentation and consumption experiences, it is important to understand the affordances and characteristics (both pros and cons) of different devices. We thus took a closer look at the wide range of "devices" and proposed six dimensions to characterize their affordances—live presentation capability, perception size, level of immersion, portability, level of isolation, and haptic feedback. We described how to tell visual data-driven stories for different devices and media with example uses of the devices in visual data-driven storytelling. We also discussed the evolution of display- and interaction-device technology as well as opportunities and challenges these new and refined devices bring to the visualization community.

## REFERENCES

Adams, E. (2004). Postmodernism and the three types of immersion. Gamasutra. http://designersnotebook.com/Columns/063_Postmodernism/063_postmodernism.htm (accessed 2/3/2017).

Bowman, D. A., Kruijff, E., LaViola, J. J., and Poupyrev, I. (2005). *3D User Interfaces*. Addison Wesley, Boston, MA.

Carmack, J. (2013). Latency mitigation strategies. https://www.twentymillisec-onds.com/post/latency-mitigation-strategies (accessed 2/3/2017).

Chandler, T., Cordeil, M., Czauderna, T., Dwyer, T., Glowacki, J., Goncu, C., Klapperstueck, M., Klein, K., Marriott, K., Schreiber, F., and Wilson, E. (2015). Immersive analytics. In *Big Data Visual Analytics (BDVA)*. IEEE, Hobart, TAS, pp. 1–8.

Cordeil, M., Dwyer, T., and Hurter, C. (2016). Immersive solutions for future air traffic control and management. In *Proceedings of the 2016 ACM Companion on Interactive Surfaces and Spaces*. ACM, Niagara Falls, ON.

Cordeil, M., Bach, B., Li, Y., Wilson, E., and Dwyer, T. (2017). A design space for spatio-data coordination: Tangible interaction devices for immersive information visualisation. In *Proceedings of Pacific Vis 2017*, pp. 46–50. Seoul, South Korea.

Dalsgaard, P. and Halskov, K. (2014). Tangible 3D tabletops. *Interactions*, 21(5), 42–47.

Ducasse, J., Brock, A., and Jouffrais, C. (2017). Accessible interactive maps for visually impaired users. In Edwige E. Pissaloux, Ramiro Velazquez (Eds) *Mobility in Visually Impaired People—Fundamentals and ICT Assistive Technologies*. Springer, Berlin.

Goncu, C. and Marriott, K. (2011). GraVVITAS: Generic multi-touch presentation of accessible graphics. In *IFIP Conference on Human-Computer Interaction*. Springer, Berlin and Heidelberg, pp. 30–48.

Goffin, P., Willett, W., Fekete, J. D., and Isenberg, P. (2014). Exploring the placement and design of word-scale visualizations. *IEEE Transactions on Visualization and Computer Graphics*, 20(12), 2291–2300.

Hettinger, L. J. and Riccio, G. E. (1992). Visually induced motion sickness in virtual environments. *Presence: Teleoperators & Virtual Environments*, 1(3), 306–310.

Huang, F. C., Chen, K., and Wetzstein, G. (2015). The light field stereoscope: Immersive computer graphics via factored near-eye light field displays with focus cues. *ACM Transactions on Graphics*, 34(4), 60.

Isenberg, P., Isenberg, T., Hesselmann, T., Lee, B., Von Zadow, U., and Tang, A. (2013). Data visualization on interactive surfaces: A research agenda. *IEEE Computer Graphics and Applications*, 33(2), 16–24.

Jakobsen, M. R., Haile, Y. S., Knudsen, S., and Hornbæk, K. (2013). Information visualization and proxemics: Design opportunities and empirical findings. *IEEE Transactions on Visualization and Computer Graphics*, 19(12), 2386–2395.

Leclaire, J. and Tabard, A. (2015). R3S. js—Towards responsive visualizations. In *Workshop on Data Exploration for Interactive Surfaces DEXIS 2015*, p. 16.

Lee, B., Riche, N. H., Isenberg, P., and Carpendale, S. (2015). More than telling a story: A closer look at the process of transforming data into visually shared stories. *IEEE Computer Graphics and Applications*, 35(5): 84–90.

Lee, B., Isenberg, P., Riche, N. H., and Carpendale, S. (2012). Beyond mouse and keyboard: Expanding design considerations for information visualization interactions. *IEEE Transactions on Visualization and Computer Graphics*, 18(12), 2689–2698.

Lee, B., Kazi, R. H., and Smith, G. (2013). SketchStory: Telling more engaging stories with data through freeform sketching. *IEEE Transactions on Visualization and Computer Graphics*, 19(12), 2416–2425.

Milgram, P. and Kishino, A. F. (1994). Taxonomy of mixed reality visual displays. *IEICE Transactions on Information and Systems*, 77(12), 1321–1329.

Munzner, T. (2014). *Visualization Analysis and Design*. CRC Press, Boca Raton, FL.

The Nielsen Company. (2014). How smartphones are changing consumers' daily routines around the globe. Nielsen Market Research. http://www.nielsen.com/us/en/insights/news/2014/how-smartphones-are-changing-consumers-daily-routines-around-the-globe.html (accessed 2/8/2017).

Roberts, J. C., Ritsos, P. D., Badam, S. K., Brodbeck, D., Kennedy, J., and Elmqvist, N. (2014). Visualization beyond the desktop—The next big thing. *IEEE Computer Graphics and Applications*, 34(6), 26–34.

Taher, F., Jansen, Y., Woodruff, J., Hardy, J., Hornbæk, K., and Alexander, J. (2017). Investigating the use of a dynamic physical bar chart for data exploration and presentation. *IEEE Transactions on Visualization and Computer Graphics*, 23(1), 451–460.

Willett, W., Jansen, Y., and Dragicevic, P. (2017). Embedded data representations. *IEEE Transactions on Visualization and Computer Graphics*, 23(1), 461–470.

# CHAPTER 7

# From Analysis to Communication

*Supporting the Lifecycle of a Story*

Fanny Chevalier

*University of Toronto*

Melanie Tory

*Tableau Research*

Bongshin Lee

*Microsoft Research*

Jarke van Wijk

*Eindhoven University of Technology*

Giuseppe Santucci

*Sapienza University of Rome*

Marian Dörk

*University of Applied Sciences Potsdam*

Jessica Hullman

*University of Washington*

## CONTENTS

## INTRODUCTION

Coinciding with advancements in Web-based visualization technology and the proliferation of mobile devices, there has been a rapid growth of novel visual data stories on the Web as well as in other formats such as infographics and data videos (Amini et al. 2015, 2017). Journalists, illustrators, developers, and analysts often work in mixed teams to develop new and richer reader experiences by introducing novel storytelling techniques and refining existing ones. A broad range of skills, disciplines, tools, and technologies comes together to support the creation of these visual-data stories.

Yet there is limited research that describes the various tools and methods that authors employ to create visual data stories, the processes they go through, and the major pain points they experience. In this chapter, we discuss the current practices as expressed by data storytellers we interviewed, give an overview of the tools that are in use, and detail opportunities for research and design. We focus on the visual data storytelling process, from conception through production, including data collection and preparation, data analysis, story development, and visual presentation. We are particularly concerned with the tools and practices that are involved in constructing a story that is driven or accompanied by data visualization.

In discussing a high-level overview of storytelling and outlining the research opportunities, Kosara and Mackinlay (2013) use a working model for how stories are constructed based on how journalists work. More recently, Lee et al. proposed a visual data storytelling process derived from the existing models in the data journalism literature. It presents an encompassing view of the storytelling process, integrating three steps—(1) finding insights, (2) turning these insights into a narrative, and (3) communicating this narrative to an audience (Lee et al. 2015). They provide a detailed description of the main roles and activities that storytellers engage in as they turn raw data into a visually shared story. We build on Lee et al.'s model, and propose a revised process, grounded in interviews with professionals who create visual data stories. As in Lee et al.'s work, we discuss the main activities, resulting artifacts, as well as roles that are involved in the storytelling process.

## UNDERSTANDING CURRENT PRACTICES: INTERVIEWS WITH DATA STORYTELLERS

While storytelling has gained substantial interest in data visualization research (Segel and Heer 2010, Hullman and Diakopoulos 2011, Hullman et al. 2013b, Kosara and Mackinlay 2013, Lee et al. 2015, Stolper et al. 2016), we are specifically interested in approaches and tools currently used in practice. We conducted interviews with nine visual data storytellers to better understand the process that they undergo to construct visual stories with data. In this section, we describe our methodology for eliciting and analyzing feedback from these professionals, and discuss major insights gained from our qualitative analysis.

### Methodology

We interviewed nine professionals (Table 7.1) including journalists, designers, and researchers who had experience in authoring visual data stories, as a single author, as part of a collaborative team, or both. We elicited responses from each professional individually using a semistructured interview protocol (Figure 7.1). Our interviews focused on the process they undergo to create a visual data story, from the ideation phase, to the execution, to the publication, and beyond. We also inquired about the different roles involved in the creative process as well as the tools a team of storytellers use to perform data analysis, design visualizations, build the narrative, document the process, and collaborate around a visual data story project. To better capture the entire process from end to end, we asked the interviewees to pick a concrete example of a project that they felt would be interesting to share. We asked them to choose any project they considered worth discussing; for example, it could be a typical project, or conversely, could be a unique project that they were particularly proud of. We asked the interviewees to describe the process with that particular example in mind, but encouraged them to deviate from that instance

FIGURE 7.1 Interview with Xaquín G.V., editor at Guardian Visuals, *The Guardian*.

whenever relevant, to comment on differences in the authoring process for other projects or atypical activities. Interviews lasted between 30–90 minutes overall; no time constraint was imposed.

All interviews were audio recorded and transcribed. Though they were semistructured to guarantee that each interview covered all aspects we were interested in, the interviews were more like informal discussions around a particular data story project than scheduled around a predefined list of questions. We therefore collected rich, qualitative information that we coded using a thematic analysis (Boyatzis 1998). Three of the chapter authors thoroughly read through all of the transcripts, noted all interesting comments and events, and roughly organized this data in categories. Here, we report the most notable insights that emerged from our analysis, starting first with an overview of the storytelling process.

Table 7.1 summarizes the interviewees' backgrounds including their profession, experience in working with data to tell stories, and the type of organization in which they currently work. Table 7.2 presents details about the project instance that was described by each of our interviewees.

## Storytelling Process

> *It's never streamlined... you go back and forth from the collection to the analysis, the analysis to the collection, and then to the sketching. And then on the sketching you realize you may be missing data and you go back to collection. So it's not a line. It's sort of like a bunch of threads that keep getting intermingled with each other.* (Xaquín G.V.)

TABLE 7.1  Summary of the Interviewees

| Participant Number | Interviewee | Profession | Experience | Organization | Organization Type |
|---|---|---|---|---|---|
| P3 | Dominikus Baur | Visualization and interaction designer | ~ 10 years | Andechs, Germany | Freelancer |
| P6 | Nicholas Diakopoulos | Data journalist and journalism research | ~ 5 years | University of Maryland College Park, MD, United States | Academia |
| P4 | Kennedy Elliott | Developer journalist | ~ 6 years | *The Washington Post* United States | Newsroom |
| P2 | Christina Elmer | Journalist | ~ 10 years | Spiegel Online GmbH Hamburg, DE | Newsroom |
| P9 | Xaquín G.V. | Editor | ~ 15 years | Guardian Visuals, *The Guardian* London, GB | Newsroom |
| P8 | Ulrike Köppen | Journalist | ~ 12 years | Bayerischer Rundfunk Munich, DE | Newsroom |
| P5 | Moritz Stefaner | Designer | ~ 10 years | Truth & Beauty Lilienthal, DE | Freelancer |
| P7 | Stefan Wehrmeyer | Computer scientist | ~ 5 years | Correctiv Berlin, DE | Newsroom |
| P1 | Benjamin Wiederkehr | Interaction and information designer | ~ 7 years | Interactive Things Zürich, CH | Digital design studio |

This quote exemplifies the experience we heard repeatedly in our interviews. While there are phases to the process of building and telling a story, there can be a lot of iteration and progression of the work that is rarely linear. Journalist Kennedy Elliott commented that *"A lot of the process is iterative and it's hard to capture."* This complexity was confirmed by the other interviews, as we found a lot of variations in the process depending on the project. Each visual data story project has its own back story: the people involved, ranging from a few individuals to a large team; the origin or initiation of the story; the data collection and wrangling process; the brainstorm phases and design—all have their specificities from one project to the next, even within the same creative team. The Editor of the

**TABLE 7.2** Projects Discussed During the Interviews

| Interviewee | Project | Team Size | Duration |
| --- | --- | --- | --- |
| Dominikus Baur | Subspotting: Mapping cell phone reception on the New York City subway. http://do.minik.us/blog/subspotting | 2 people | ~ 1.5 years (side project) |
| Nicholas Diakopoulos | An article describing search engine bias around different politicians. (Trielli et al. 2016) | 3 people | ~ 3 months (not full time) |
| Kennedy Elliott | Investigation of police shootings. Created a database of all the police reports concerning someone being shot (back end for reporters and investigators) and front end database for the public. https://www.washingtonpost.com/graphics/national/police-shootings-2016/ | ~ 200 people, organized in subteams | ~ 2 years, still ongoing (not full time) |
| Christina Elmer | Comparison of nursing house prices when you sell your house on the market, and the money you have to be able to spend on nursing homes. http://www.spiegel.de/wirtschaft/betongold-die-interaktive-deutschlandkarte-a-1047006.html | | ~ 2 months (not full time) |
| Xaquín G.V. | The drop in imports in China was going to bring down some countries or was going to have a huge effect on other countries. https://www.theguardian.com/world/ng-interactive/2015/aug/26/china-economic-slowdown-world-imports | The whole newsroom | ~ 6 weeks |
| Ulrike Köppen | Schnee von morgen: A multimedium data story (interactive platform online and documentary film for TV) about how the climate change in the Bavarian Alps over the last decades influenced the local economic situation and tourism industry. http://schnee-von-morgen.br.de/ | 6 people | ~ 2 months |
| Moritz Stefaner | Where the wild bees are: one page static visualization investigating the change of the bee population and bee pollinator interaction over 120 years in a certain region within the United States (Jabr 2013) Documented design process: http://well-formed-data.net/archives/972/where-the-wild-bees-are | 2 people | ~ 6 weeks (not full time) |
| Stefan Wehrmeyer | Hospitals get more funding for psychiatric beds than regular beds so there is encouragement to keep the patients there, even though home care is better for them. https://correctiv.org/recherchen/stories/2015/09/26/der-psychiatrie-skandal | 3 people | ~ 1 month (not full time) |
| Benjamin Wiederkehr | Infanticide and abortion of girls in India. Girls are not reported as having been born. Looking for data to demonstrate that the number of girls is not what should be expected. https://lab. interactivethings.com/unwanted | The whole newsroom | ~ 3 weeks/250 hours of labor (not full time) |

Visuals team at *The Guardian*, Xaquín G.V., commented towards the end of the interview, "*We have such a huge variety of projects [...] the process that I described is kind of like an ideal. Sometimes it's messier.*"

Nevertheless, we found consistency across all projects in terms of their high level phases, which also aligned with previously proposed models (Kosara and Mackinlay 2013, Lee et al. 2015). Lee et al.'s (2015) model of the storytelling process comprised three main phases: *explore data, make a story,* and *tell a story,* which we refined and further developed to reflect the insights from the interviews (Figure 7.2). We discuss each phase in detail. For simplicity, we present them in order, but the reality involves many back-and-forth iterations between different phases.

***Exploring data*** involves the set of activities centered around investigating and analyzing data. Lee et al. (2015) suggested that the analysis is necessarily exploratory in nature, where the data analysts are in search of an interesting subject matter to make a story (Figure 7.3). This was backed up by our interviews for some projects. Moritz Stefaner, who works as a freelance designer, said that sometimes clients come to him and "*just say they want to do something interesting with that data, and then [he has] to tell them what the interesting thing is they can do with the data.*" However, our interviews revealed that the story outcome was often known (or correctly hypothesized) before the data was in the hands of the storytellers;

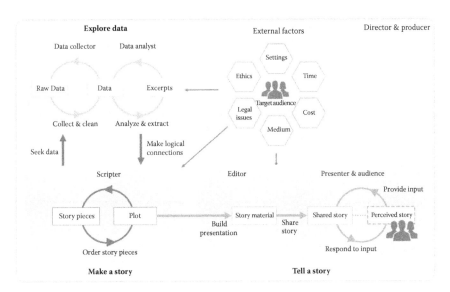

FIGURE 7.2  The storytelling process from the story idea to visually shared stories.

(a)

(b)

FIGURE 7.3 Illustration of the exploring data phase in a newsroom. Data exploration is conducted using various artifacts and devices (boards, post-its, computers, etc.).

in this case, data analysis was confirmatory, performed to check facts and produce accurate figures to substantiate the narrative.

Regardless of the origin of the story (we discuss this point later), the exploring data phase consists of extracting relevant data excerpts from the raw material, i.e., simple data facts or more complex data insights that can be used to make and support a narrative. Even when the main story outcome is known a priori, this data analysis process is still necessary to verify the story's claims and identify the most pertinent data fragments to get a point across through a visual story. Moritz Stefaner, who worked on the redesign of a figure from a study done by researchers (Figure 7.4), insisted that even though *"the data analysis was done [...] the first step is always understanding the material [i.e.,] the completeness, distribution, and also the texture of the data."* Furthermore, the designer had to do a

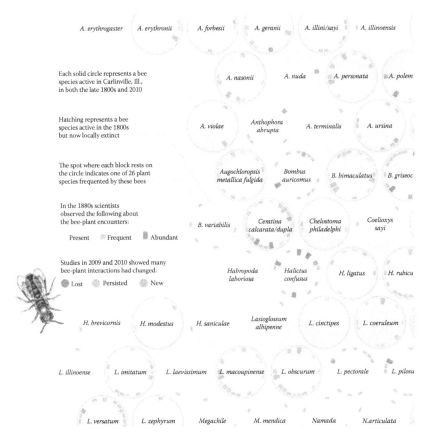

FIGURE 7.4    Excerpt of a static visual data story by Moritz Stefaner, designed for an article in the *Scientific American* (Jabr 2013).

lot of data exploration "*because [he] had a whole set of findings, but [he] had to identify what are the most interesting ones for a general audience*" (Figure 7.5).

An interesting case of data exploration that was mentioned by Ulrike Köpper concerned the underlying analytical reasoning behind data exploration, and how it is sometimes necessary to "*find new ways of looking into the data.*" In the Schnee von morgen project, the investigators naturally aggregated economical and climate data by administrative region, which made it "*really hard to see a story.*" The journalist explained how they "*came back to how to tell the story about this data,*" and found that their analytical reasoning was flawed: climate data is not ruled by territorial divisions. The team eventually devised an entirely new analytical

(a)

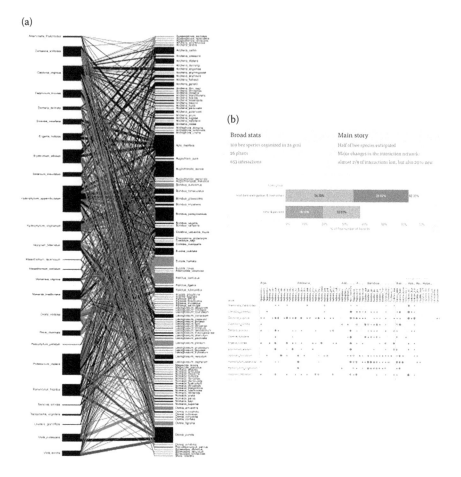

FIGURE 7.5 (a) Original chart from the research article for the study (Burkle et al. 2013). (b) Moritz Stefaner's result of initial data extraction and analysis.

strategy, *"making [their] own regions because every story perhaps has its own geographic idea behind it."*

To do analysis, data must be available to the data analyst, and this data must be clean and in a usable format. Lee et al. (2015) briefly mention data preparation, i.e., collection, cleaning, and integrity-checking activities associated with the exploring data phase. Our interviews revealed that such data-preparation activities often constitute a significant portion of the work.

*"Finding data can be a pain,"* said Xaquín G.V. This statement was echoed by Stefan Wehrmeyer who commented that *"it's always hard to come by data."* Since *"it is impossible to just say you demand the data,*

*and two days later, it will come back,"* (Benjamin Wiederkehr) storytellers sometimes predict the data they will have available and start the design process in parallel, hoping to find, and be granted access to data relevant to the story. Benjamin Wiederkehr related a case where they waited for the raw data *"until the last day,"* forcing the team to *"find a way to tell a story of a dataset without having the raw numbers."* The designer, Moritz Stefaner, also mentioned that *"sometimes you have the communication goals, but the data you have doesn't get you there."* So, even when the raw data is handy, *"sometimes you have to go for more data to put that other dataset in context"* (Xaquín G.V.).

Raw source data is oftentimes obtained from public or private sources, though Ulrike Köpper pointed to the difficulties of obtaining private data (in her case, on investment), as they found *"it was absolutely impossible to have people collaborate with you and share private data concerning money because of taxes and related financial stuff."* When raw data is unavailable, storytellers sometimes reverse engineer existing charts, a practice that Benjamin Wiederkehr says *"is not entirely accurate"* but sometimes is *"as good as [one] can do"* to collect the data. Data can also sometimes be created by the project team itself. For instance, Dominikus Baur, who worked on mapping the cell phone signal strength over the New York subway network explained that *"first [the team] needed metadata about the lines themselves,"* i.e., all of the stations and their geographical coordinates which they obtained from the transportation authority. Then the team wrote a data logger, which they used for data collection by riding the subway in person (Figure 7.6a). In this specific project, preparing data involved planning for data collection based on existing (proprietary) metadata, developing a custom application to record the data, recording the data by riding the subway, and further cleaning and wrangling activities (Figure 7.6b).

Kennedy Elliott's project also entailed a considerable amount of data preparation work. In order to provide empirical data about people killed by the police in the United States, her team at *The Washington Post* collected all of the police reports concerning someone being shot and created a comprehensive database, part of which is publicly accessible.* This process took about 2 years to complete, but now constitutes a rich resource that allowed the journalists to tell not just one but dozens of stories.

---

* https://www.washingtonpost.com/graphics/national/police-shootings-2016/.

(a)

(b)

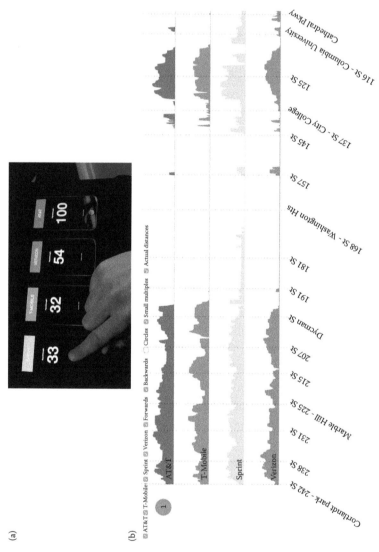

FIGURE 7.6 (a) Custom data logger used to collect cell signal strength for the Subspotting project. (b) Custom debug visualization developed for data cleaning.

When data is obtained from a third party, cleaning and reformatting is usually necessary. For instance, Stefan Wehrmeyer obtained two datasets about hospitals from government sources, one for 2003 and the other for 2013. He reported difficulties matching the hospitals between the 2 years because some of the hospital names had changed. The team eventually "*dropped the comparison of the years on an individual level*" due to this problem. Ulrike Köpper explained that the team "*had to bring some intelligence*" in the data processing of climate data, as the data collector "*had to become a true expert on snow depth, temperature and all the mathematics behind that*" to properly prepare the data for analysis.

Data may be incomplete and difficult to find, and when available, it can also be corrupted. Most of our interviewees commented that the analysis and exploration phase always starts with checking data integrity to identify inconsistencies and verify completeness. Moritz Stefaner mentioned that one "*shouldn't take data as given, but always assume that there is a flaw in the data.*" He advocated for the team to systematically track down inconsistencies and, whenever possible, check basic facts about the data. Journalist Stefan Wehrmeyer identified inconsistencies even in raw data from a government statistics office, "*There's a total column for all beds [in the data spreadsheet]. If you sum the individual beds, you get most of the times more than the total.*" He further commented, "*you have to take everything with a caveat,*" and reported contacting hospitals individually to point out the anomaly and ask for corrections. Cristina Elmer reported a similar issue when she "*found data points that weren't quite clean*" in the nursing homes data set her team obtained from a professional agency, further commenting that "*this is a regular data problem that you have with corrupt data.*" Like Stefan Wehrmeyer, she too phoned nursing homes to find out what corrections to apply; she did not consider this out of the ordinary, stating that the experience was "*usual stuff*" when working on data storytelling.

The data exploration phase is often associated with sketching, plotting, and charting, and therefore is often intermingled with exploration of visual externalizations of the data, though the specification of visual representations is not necessarily set at this stage.

*Making a story* consists of selecting and assembling the identified data excerpts in a sequence that forms a coherent and compelling narrative (Figure 7.7). Starting with a collection of key story points and data excerpts resulting from the data exploration phase, the storytellers still have "*to break [this information] down, and prioritize, and bring it into sequence,*"

FIGURE 7.7    Figuring out the best ways to form a compelling narrative by testing visual alternatives in the making a story phase.

as Moritz Stefaner described. Our interviews confirmed that the sequence plays a critical role in a story. The main activities in this phase include setting the stage, ordering, establishing logical connections and important context, developing flow, formulating a message, and creating the arc and denouement of the storyline or plot. These activities often require multiple iterations, making it necessary to go back to the data exploration phase to gather not only more excerpts but also more data (Gratzl et al. 2016). For instance, Ulrike Köpper reported that for their project, the team started by looking at economics data to gain initial insights. When they found facts about some regions, *"they could look for people who could tell how this happened [...] and why it happened."* After she extracted major insights to make a story about abortion of girls in India, the designer in Benjamin Wiederkehr's team came up with the idea to combine the quantitative information with qualitative data as a way to reinforce the narrative. She collected, from documentaries, a series of personal statements and personal stories from people affected by the problem, which implied further iterations of *"what quotes would come in, what descriptive text would come in, and then what quantitative information would come in."*

The result of the making a story phase is the story plot, which describes how the story pieces are related (e.g., in time, cause and effect, patterns, etc.) and what they mean in an overall context.

***Telling a story*** consists of materializing the story plot and delivering the story to the audience. Three main activities in this phase are building a presentation, sharing the story using the story material, and finally, receiving and handling feedback from the audience if available (Figure 7.8). During the building stage, plot and story pieces are turned into story material. Each piece of abstract content is materialized through the development of visual representations, interactions, animations, annotations, or narration. For example, story material could consist of one or more

(a)

(b)

FIGURE 7.8   Interactive data story (a) is consumed by an audience (b) in the telling a story phase.

visualizations assembled in a slide deck, a video with narration, an infographic presented on a poster, or a demo planned with an interactive system for the live presentation. These visual elements—the *"visual vehicle,"* as described by Moritz Stefaner—were mentioned as essential criteria for powerful communication.

The story material turns into a shared story once it is delivered to an audience. Ultimately, the perceived story is what the audience understands through the storytelling experience. One important aspect is about how story elements may be used to engage the audience. For example, Xaquín G.V. commented about *The Guardian's* extra efforts to convey an emotional connection: *"there was something we were missing— the almost emotional connection between what we were trying to say, and what we were visualizing."* In their ultimate design, the downward dragging interaction of a country was mapped to the drop of its imports. Furthermore, when people drag China down to simulate its imports dropping, the effect propagates dragging the rest down with it; he commented that this provided *"such an immediate read of the story, that [...] you got a very emotional reaction from a reader."*

Similarly, the designer of the Unwanted Project engaged the audience emotionally by integrating poignant quotes from people who were directly exposed and confronted with abortion in India. She also consciously employed a distinct visual aesthetic, to *"give some character [while] trying to be sensible to the topic [and] not over dramatizing"* (Figure 7.9).

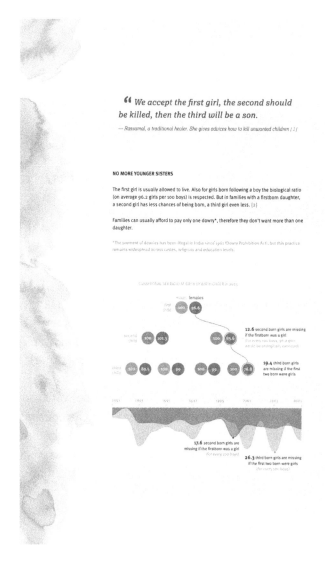

FIGURE 7.9 Screenshots from "Unwanted," by Interactive Design. The visual design and quotes from people facing the girl abortion issue in India aim to create a strong emotional connection. Source: https://lab.interactivethings.com/unwanted/#/story-01.

In current practice, making a story is often intertwined with building the presentation, as also acknowledged by Lee et al. (2015). For example, making a presentation which involves sequencing specific visual representations builds a plot. However, by making the distinction explicit in the process, we can capture the different challenges storytellers face within these two components. For example, making a story and building story material require different skill sets.

## Origin of the Story

A significant element that has been missed in prior discussions around the process pertains to the origin of the story. It was assumed that a story typically emerged during the extract and analyze data phase of the process, that is, the story usually emerged from the data: *"sometimes the dataset is a defined thing, because you just found this dataset; you think that there might be a story there,"* said Xaquín G.V. But through our interviews, we learned that the storytelling process is often initiated by a *story idea* rather than insights from exploratory analysis. For most of the projects that were discussed during the interviews, someone (often a journalist) had an idea of the story to tell coming out of personal experience, interest, or journalistic reporting, and the work consisted of seeking some data to investigate the issue and see if further evidence could be found to backup the intended story.

Alternatively, the story may already exist in a written form, and is recast as a visual story. This was the case of one of our interviewees, Moritz Stefaner, who was commissioned by the popular science magazine *Scientific American* to build a visual story—an illustration to be printed in the magazine (Jabr 2013)—from a published research article investigating the change of bee population and bee pollinator interaction over 120 years in a certain region of the United States (Burkle et al. 2013). In this project, the storyline and overall message were already known in advance, and the raw data, as well as the original researchers' insights, were available to the designer. However, to translate the main message of the scientific research article, the designer still performed his own data exploration, both to check that the data was correct and to further confirm with the researchers that he properly understood the core aspects to be communicated through the visual story artifact (Figure 7.6b).

## Roles

With the diverse activities comprising the process of creating a visual data story, there are also various roles and skill sets distributed among the coauthors such as artists, designers, and developers (see Chapter 8, entitled "Organizing the Work of Data-Driven Visual Storytelling," for more details on how these roles and skill sets are often constituted into functioning teams). The *data collector* gathers the data and makes it usable through a data wrangling process. The *data analyst* examines and explores the data to identify meaningful insights or excerpts. A *scripter* builds the overall plot using these excerpts. The *editor* prepares the story material for the target media based on the plot. The *presenter* delivers the story to the audience members, who consume the story and provide feedback. The *director* overlooks the project and ensures the timely completion of the project. Coordination to ensure that everyone works towards the same direction is especially important for a large project involving a complex team, where subteams and roles may evolve through the course of the project. For example, the police shooting database project involved about 200 people over 2 years. In addition, in some cases, the *producer* (or client) initiates the creation of a visual data story by hiring storytellers. Many of these roles (data analyst, scripter, editor, and presenter) were also identified in related work (Lee et al. 2015). Note that roles are not synonymous with individuals: each contributor may take on one or many different roles. Ulrike Köppen insisted on the interdisciplinary nature of data storytelling and argued that for the team to function well, *"everyone has to know what the other needs for their work,"* and so each member of the team, regardless of their role(s), *"has to know a little bit about what coding means, what graphics means, what design means so that [one's] knowledge can serve the others to do better work."*

## Storytelling Constraints

External factors can substantially influence the creation process. Lee et al. (2015) discussed three external factors: (1) the target audience at the receiving end of the shared story (see Chapter 9 entitled "Communicating Data to an Audience"), (2) the setting in which a story is told, and (3) the medium that conveys the story. Our interviews revealed two additional important external factors that slowed down, or even hindered, the creative process: *time and cost constraints* and *legal and ethical implications*.

Many of our interviewees mentioned the time pressure to deliver stories, a typical constraint in journalistic circles: *"We basically have time constraints and resource constraints because we're doing too much at the same time"* (Stefan Wehrmeyer). Kennedy Elliott noted, *"We can't always be very iterative because we're on a tight load deadline,"* and expressed the usual lack of time for iterative story design, saying, *"This one was exceptionally iterative, like I did three different versions."* Another interviewee, Moritz Stefaner, said, *"In [the] magazine world you have only a few days of actually working time."* This time pressure often meant that storytellers could not fully complete what they envisioned initially. For example, journalists said that they, *"didn't have time to build the sharing button for every view, every region,"* (Christina Elmer) or, *"to make the graphic switch focus between countries"* (Xaquín G.V.). Cost can also be limiting. In addition to labor costs, storytellers may have to buy data from private sources as mentioned by two of our interviewees, Christina Elmer and Stefan Wehrmeyer. Dominikus Baur told us that he and his colleagues decided to pay the subway company to get the license *"for the visual identity"* because they wanted to associate their work with the official transportation company to build trust from the audience. Legal implications, e.g., copyright, as well as ethical issues can also impact the process and the resulting visual data story. For instance, Kennedy Elliot, the journalist involved in the project about deaths caused by police, said that in many occasions, they wondered, *"Do we publish this? Can we disclose that?"* (See Chapter 10 entitled "Ethics in Data-Driven Visual Storytelling".)

## TOOL LANDSCAPE

The interviews revealed that a wide variety of systems, tools, and technologies are used for producing visual data stories. In the following, we aim to provide an overview of the landscape. The goal is not to provide recommendations on what to use for what purpose, but to provide insights on patterns and trends. The landscape is dynamic: this overview presents tools and systems that were in use in 2016, and very probably new tools will emerge and others will lose popularity. However, most important are not the names of the tools but their features, as we expect these to remain relevant.

There are several factors that contribute to the variety of tools in use. First, as described in the previous section, there are many different steps in the process, ranging from data acquisition and analysis up to story scripting and delivery to an audience. People involved have different expertise, skills, backgrounds, and preferences: while a data analyst may be familiar with R or Python, a graphic designer might use Illustrator as their standard tool. To present the resulting story and visualization, a wide range of media can be used, including print, videos, websites, and apps, each with their own highly specialized technology stack. Companies and institutes often have de facto standards for preferred tools, and for instance, a local data base management system (DBMS) or content management system (CMS) can impose considerable constraints on the production of a data story. The quality level required varies: sometimes basic statistics and graphics suffice, in other cases a sophisticated analysis or careful tuning is required. In addition, the data to be shown varies, including numeric, geographic, network, and text data. For each data type, dedicated tools are available for analysis and visualization. As a result of all this, in practice, a variety of tools are used in combination. Table 7.3 provides an overview, listing all the tools and formats that were reported in the interviews. We classified these based on the experience of the interviewees, but acknowledge that many of the tools can and are used for broader purposes as well. We classified the tools according to the following characteristics:

- The type of the tool, describing its primary purpose, i.e., not considering the storytelling lifecycle.

- The specificity of the tool, in terms of how many activities it is used for. *High* means that the tool is very specific and covers only one activity; *low* means that the tool may be used in several different activities.

- The abstraction of the tool, i.e., the power of the tool functionalities. A general purpose programming language is classified *low* because one single instruction has little power, a tool producing interactive images from data is classified *high*.

- The type of projects that used/mentioned it.

- Its type of license (free/commercial).

- The skill level that is required to use it.

- The phases of the storytelling process in which the tool has been used (see Figure 7.1).

- The roles of the people using the tool (see Figure 7.1).

It is interesting to note that more than 40 different tools and formats were used in the projects and that about half of the tools have been used by only one project. Moreover, general purpose programming languages like Python and JavaScript are used in all projects; specifically, five projects

TABLE 7.3  Tools and Formats Used in the Projects

| Name | Url | Type | Specificity | Abstraction | Projects | License | Skill | Phases | Roles |
|---|---|---|---|---|---|---|---|---|---|
| Amazon Mechanical Turk | www.mturk.com/mturk/welcome | Crowdsourcing | High | Medium/high | P6 | Commercial | High | Explore data—make a story | Data collector—scripter |
| Basecamp | basecamp.com | Collaboration | Medium | Medium | P1 | Commercial | Medium | All | All |
| Chrome Inspector | developers.google.com/web/tools/chrome-devtools/ | SW tool | Low | Low | P3 | Free | High | Explore data | Data collector—data analyst |
| Chrome Console | developers.google.com/web/tools/chrome-devtools/ | SW tool | Low | Low | P3 | Free | High | Explore data | Data collector—data analyst |
| CSV or TSV | en.wikipedia.org/wiki/Comma-separated_values | Standard exchange format | Low | Low/medium | P3, P5, P7 | Free | Low | Explore data | Data collector—data analyst |
| D3 | d3js.org | Programming language for web visualizations | Medium/High | Low | P1, P3, P4, P5, P9 | Free | High | Explore data—make a story | Data collector—data analyst—scripter |
| Dropbox | dropbox.com | File sharing/collaboration | Low | Medium | P1, P3 | Free/Commercial | Low | All | All |
| e-mail | en.wikipedia.org/wiki/Email | Communication | Medium | Medium | P1, P2, P3, P5, P6, P7 | Free | Low | All | All |
| Excel/Google sheets | en.wikipedia.org/wiki/Google_Docs,_Sheets,_and_Slides | Spreadsheet | Low | Medium | P1, P2, P5, P7, P9 | Free | Medium | Explore Data—Make A Story | Data Collector—Data Analyst—Scripter |

(Continued)

TABLE 7.3 (*Continued*)  Tools and Formats Used in the Projects

| Name | Url | Type | Specificity | Abstraction | Projects | License | Skill | Phases | Roles |
|---|---|---|---|---|---|---|---|---|---|
| GeoJSON | geojosn.org | Geo format | Medium | Low/medium | P7 | — | Medium | Explore data | Data collector—data analyst |
| Gephi | gephi.org | Graph drawing tool | High | High | P5 | Free | Medium/high | Explore Data | Data analyst |
| GID | www.gidhome.com | PrePost numerical simulation processing | Medium | Medium/High | P5 | Commercial | High | Explore data | Data analyst |
| GitHub | github.com | SW development | High | Medium | P3, P7 | Free | Medium | All | All |
| Google analytics | en.wikipedia.org/wiki/Google_Analytics | Web pages statistics | High | Low | P2, P3 | Free | Low | Tell a story | Presenter and audience |
| Google docs | en.wikipedia.org/wiki/Google_Docs,_Sheets,_and_Slides | Shared documents | Medium | Medium | P6, P7 | Free | Low | All | All |
| Google hangouts | hangouts.google.com | Communication | High | Medium | P5 | Free | Low | All | All |
| Harvest | www.getharvest.com | Management | High | Medium | P1 | Commercial | Medium | All | All |
| High Charts | www.higharts.com | Visual analysis | Medium | High | P2 | Commercial | Medium/High | Explore data—make a story | Data analyst—scripter |
| Illustrator (Adobe) | www.adobe.com/products/illustrator.html | Vector art tool | High | Medium | P1, P4, P5, P6, P7 | Commercial | Medium/High | Tell a story | Editor |
| Infobase | www.infobaseindex.com/index.php | Data source | High | Medium | P1 | Free | Low | Explore data | Data collector—data analyst |

(*Continued*)

**TABLE 7.3 (*Continued*)** Tools and Formats Used in the Projects

| Name | Url | Type | Specificity | Abstraction | Projects | License | Skill | Phases | Roles |
|---|---|---|---|---|---|---|---|---|---|
| Infogram | infogram | On line charting | Medium | High | P6 | Commercial | Medium | Explore data—make a story | Data analyst—scripter |
| JavaScript | en.wikipedia.org/wiki/JavaScript | Programming language | Low | LOW | P1, P2, P3, P4, P5 | Free | High | All | All |
| Json | en.wikipedia.org/wiki/JSON | Format | Low | Low/medium | P5, P7 | — | Medium | Explore data | Data analyst |
| Jupyter notebook | Jupyter.org | Shared documents (with equation/code/viz) | Low | High | P7 | Free | Medium/High | Explore data—make A story | Data collector—data analyst—scripter |
| Keynote | App store | Presentation | Medium | Medium/high | P1 | Commercial | Low/medium | Make A story | Scripter |
| Mapbox | www.mapbox.com | Map visualization | Medium | High | P2 | Commercial | Medium/high | Explore data—make a story | Data analyst—Scripter—editor |
| MySQL | www.mysql.com | DBMS | High | Medium | P6 | Free/commercial | Medium | Explore data | Data collector—data analyst |
| Objective C | App store | Programming language | Low | Low | P3 | Free | Medium/high | Explore data | Data collector—data analyst |
| OpenNames.org | Closed: https://github.com/pudo/nomenklatura/wiki/OpenNames.orgShutdownNotice | Matching names data source | High | Low/medium | P7 | Closed | — | Explore data | Data collector |

(*Continued*)

TABLE 7.3 (*Continued*) Tools and Formats Used in the Projects

| Name | Url | Type | Specificity | Abstraction | Projects | License | Skill | Phases | Roles |
|---|---|---|---|---|---|---|---|---|---|
| OpenRefine | openrefine.org | Tool for messy data | High | Medium | P5, P7 | Free | Medium | Explore data | Data collector |
| Paper/ whiteboard | — | Collaboration | Medium | Medium/ high | P1, P5, P9 | — | — | Make a story | Scripter |
| Pivot animator | pivotanimator.net/ Download.php | Animation | High | Medium/ high | P7 | Free | Medium/ high | Tell a story | Editor |
| Pricing scripts | fillz.zendesk.com/ hc/enus/ sections/201087399 SamplePricing Scripts | Scraping data | High | Low | P2 | Free | Medium | Explore data | Data collector |
| Python | www.python.org | Programming language | Low | Low | P3, P4, P5, P6, P7 | Free | Medium/ high | Explore data | Data collector— data analyst |
| R | www.r-project.org | Programming language | Low | Low | P4, P6 | Free | High | Explore data | Data analyst |
| Sketch | www.sketchapp.com | Apple tool for digital design | Medium | Medium/ High | P4 | Commercial | Medium/ high | Tell a story | Editor |
| Sublime text | www.sublimetext. com | Text editor | Medium | Low | P2, P3 | Commercial | Low/ medium | Explore data | Data collector— data analyst |
| SVG | Wikipedia.org/wiki/ Scalable Vector Graphics | Vector grapics standard format | Medium | Low | P1, P5, P7 | Free | Medium | Tell a story | Editor |
| Tableau | www.tableau.com | Visual analysis | Medium | High | P1, P5, P6 | Commercial | Medium/ high | Explore data—make a story | Data analyst— Scripter |

used Python, five projects used JavaScript, and five projects used both. That makes evident that tools with a high level of abstraction, aiming to provide a complete solution and shielding users from details, often fall short, and that often one has to rely on low-level coding.

*Exploring data* requires functionality for data acquisition, cleaning and conversion, analysis, and visualization. From a tooling landscape point of view, two approaches can be distinguished. One is a bottom-up approach. One can use a programming language (Java, C++) or scripting language (JavaScript, Python) to perform all these tasks from scratch. R also fits in this category as a programming language defined to support statistical computation in particular. Flexibility is optimal, but the downsides, especially effort and expertise needed, are clear. This can be compensated by using libraries, plug-ins, and frameworks, reusing expertise and efforts from the community. Functionality to produce graphics is provided by technologies like OpenGL and SVG; on top of these more-focused frameworks are available (like D3 for data visualization); on top of which yet more-focused modules are available, for instance, to produce histograms or other standard visualizations. Reuse can be highly efficient, but finding one's way in all of the available resources is a skill in itself. Scripting languages have become very popular as they facilitate rapid exploration and prototyping. To increase productivity further, dedicated environments to quickly enter code, see results, etc., have become very popular, with Jupyter Notebook and Zeppelin as great examples. Data scientists are fluent in using these tools, however, journalists and graphics designers often are not.

The top-down approach to tooling for exploration is based on end-user applications that aim to support one (smaller or larger) task. Examples of such tools are those for cleaning and conversion (such as OpenRefine and Trifacta Wrangler), statistical analysis (like SAS), and interactive visual analysis (including Tableau and Spotfire). The lines are blurry here as statistical analysis tools often provide visualizations, and visualization tools provide statistics. To address the needs of users, more and more functionality is added, either built-in (leading to many options and feature creep), or more by providing programming options. The most popular tool for data exploration is probably the spreadsheet (Microsoft Excel, Google Sheets), where this pattern is visible. Spreadsheets start from a simple metaphor but provide substantial additional functionality, such as options for business graphics and scripting; as a result, they can be employed for surprisingly complex tasks. But, for many tasks and data types, spreadsheets fall short and more specialized tools are needed, for instance, to analyze and visualize networks.

Highly sophisticated and professional end-user tools can be costly, both in terms of commercial costs and the effort needed to acquire a useful level of proficiency. As a reaction, simpler end-user tools are developed with a clear and crisp functionality. Some examples are infogr.am, High Charts, and DataWrapper, which support rapid development of interactive Web-based business graphics. More focused on geovisualization, services like Mapbox, CARTO, and Google Maps allow rich interactive representations of geographical data sets on various Web-enabled devices. Specializing on network data, Gephi provides advanced graph analysis and drawing methods.

For *making* and *telling* the final story, a variety of tools are also used. The data exploration phase can lead to visuals that can be used directly, or have to be augmented (using for instance Illustrator) or converted to another format. None of the interviewees mentioned the use of dedicated storytelling tools. Instead, they used mainstream generic tools for writing text and implementing interactive Web applications. To facilitate the use of various tools, allowing the import and export data in standard formats is important. For data, CSV and JSON are often used, text is exchanged via RTF, and images are exchanged using SVG (vector) or PNG (raster). Proprietary formats of popular vendor-specific tools like Microsoft Word or Adobe Illustrator are also used. However, it is worth noting that the practical usage of such standards is not always straightforward: Stefan Wehrmeyer reported that he *"tried to export it to CSV files and it was so hard..."* and a lot of activities are performed manually. Dominikus Baur said, *"So, a lot happened manually, actually, [for instance,] in Sublime Text or Atom. And most of the data cleaning I then did in the Chrome Inspector."*

A particular problem with the use of manual tools is that new data demands additional labor to recreate visualizations. Manual interventions in the data processing or added annotation during the final stages have to be redone, leading to extra work and opportunities for the introduction of mistakes. Benjamin Wiederkehr gave a vivid description of a typical project using Tableau for exploration, JavaScript/D3 for producing visuals, and Illustrator for the design. He reported that dealing with changes in the data, *"eats a lot of time unnecessarily right now."* An approach where the complete pipeline is scripted can deal with this, but often lacks the flexibility and visual appeal of manual creation. A more complex problem is that new data might also lead to changes in the story itself—an issue that cannot be addressed automatically and requires proper judgment of the editor and domain experts. Moreover, interviews made clear that different tools may be used in each stage and the same tool is often used in different phases. As an example, Illustrator is used for final storytelling and polishing as well in the conception stage as reported by Kennedy Elliot, *"So initially just Illustrator to mock things up, and pad and paper to sketch things and get ideas."*

The interviews made it clear that productions of visual data stories are often *team projects*. Hence, besides the steps centering around storytelling itself, collaboration as well as project management support are important and have to be addressed (see Chapter 8 entitled "Organizing the Work of Data-Driven Visual Storytelling"). For communication, standard solutions are used, including face-to-face meetings with whiteboards, video conferences like Google Hangouts, email, and file sharing solutions like Dropbox. Version management is important to look back at discussions and monitor progress. A simple solution is to log everything in one document, but also standard version-management solutions from software engineering (like GitHub) can be used. Well-organized design studios document and log all their results, progress, and agreements systematically such that other people can take up work if something happens to somebody. At Interactive Things, the company of Benjamin Wiederkehr, Basecamp is used for progress management, and Harvest as time tracker: *"we use Harvest as a time tracker, where you basically just track how many hours you invest in which activity."* Furthermore, they take story whiteboard photos after every important discussion.

A final process concerning visual data storytelling is measuring the impact on audiences. None of the interviewees had a systematic process for understanding impact, though they reported scanning social media and using reporting tools like Google Analytics. Also, impact can be evaluated in various ways for different audiences. Benjamin Wiederkehr remarked that for their project, the appreciative responses of organizations that were studying the topic of the story were important.

## DIRECTIONS FOR RESEARCH AND DESIGN

Our interviews of storytellers and the survey of the tool landscape point to opportunities and challenges for future research and development related to visual data storytelling.

Our interviewees consistently reported several pain points throughout the process of creating a visual data story. These pain points result in part from the lack of appropriate tools to support a process from end to end. Different phases of the process are supported by different tools, and these tools are not integrated. All had managed to overcome these difficulties to create visual stories, but wished they had more appropriate tools to support their workflow. An open question is whether the tool variety we observed is inevitable, or if a single integrative environment that supports all of the needed functionality with flexibly and extensively, might be achievable. We describe opportunities for future work according to the phases of the storytelling cycle, and outline challenges cutting across multiple or all phases.

**Data wrangling and integrity checking**, which typically occur in the exploring data phase, are important areas for future work. *"If you cannot program at all it's a huge barrier because then you cannot reshape the data,"* said Moritz Stefaner. Tools like Wrangler (Kandel et al. 2011) aim to make data reshaping less programmatic through graphical user interfaces. However, tasks like checking data integrity remain challenging for multiple reasons. Oftentimes, natural language processing is required to integrate data from differing sources. As Stefan Wehrmeyer described, *"I tried to match [the hospitals] with OpenRefine, but that didn't quite work out."* As the size of a data set grows, the task of manually specifying matching criteria between variable names or other string data typically grows as well. Machine-learning approaches may help scale large data set integration tasks, but when the final result of integration is a visualization-based story, the tolerance for errors in integration is low. Further development of tools like Wrangler that can make sophisticated reshaping and integrity checks accessible to nonprogrammers are needed. In particular, the development of interfaces that enable nonprogrammers to use and understand classification and other machine-learning techniques that can greatly increase their efficiency in data preparation for visualization-based stories is an area for future research.

Several interviewees mentioned the difficulty of finding data in the exploring data phase. While data search engines are available online through some portals,* metadata is frequently lacking, and data integrity checks are left to the user rather than part of the typical process when publicly posting data. Visualization-based storytelling will benefit from **better tools for identifying appropriate open data given an information need or topic**. There is also an opportunity to better understand the process by

---

* www.data.gov; www.worldbank.org.

which stories emerge. As multiple interviewees described, story ideas often exist before the story creation starts, such that creation may be described as a process of finding a persuasive way to visually tell a story more so than discovering a new story per se. While some research in information visualization has examined the rhetorical process by which stories are created (Hullman and Diakopoulos 2011), a deeper understanding of how designers find data and other story material, and how to support them in doing so more effectively, is still to be gained.

The making a story phase brings still more opportunities for tools and research. The impact of sequencing decisions, mentioned by several interviewees, has been the focus of empirical study (Hullman et al. 2013b). By modeling what makes a sequence effective for a given story, and what types of sequences exist, it is possible to develop cognitive cost models and other structural models that could be instantiated in tools. For example, a tool might recognize when a storyteller has created a set of visualizations that would be appropriate for a particular type of story based on the underlying data operations. Suggesting and otherwise **supporting effective sequencing of narrative data representations** remains an exciting area for future work. Given that data stories often involve multiple related representations, other opportunities include tools to help designers create better sets of visualizations through support for design consistency across views (Qu and Hullman 2016) or better aids for conveying how data across views is related (Knudson and Carpendale 2016).

Also relevant in the making a story phase is the potential for the chosen story to be well-received by the target audience. Currently, few tools aim to support **rapid evaluation of story ideas or prototype stories**. However, the existence of crowdsourcing platforms like Mechanical Turk make it relatively easy to obtain multiple perspectives on whether a story works. Future tools might focus on how crowdsourcing and other online populations could be made accessible to nonprogrammers to help a storyteller evaluate the message that forms the basis of a visualization-based story, such as by providing templates for gathering feedback and integrating the results across multiple workers. Some recent research using crowds for visual design feedback provides promising results about the potential for such feedback to lead to more effective visualizations (Luther et al. 2014, Hearst et al. 2016). There is also room for the development of further design frameworks, such as the Five Design-Sheet Method (Roberts et al. 2016), that designers can use to systematically evaluate their own emerging story. For example, a design framework might evaluate a story idea before publication by helping the designer formulate the communication objectives of the design, such as by listing the messages that an audience member should come away with after using the design. The degree to which the design supports these objectives could then be quantitatively assessed through an early-stage evaluation with a crowd.

**Bridging and porting stories between formats** comprises a large part of the telling a story phase, when the story begins to materialize. Given that storytellers often must rely on different tools with different formats at different phases in the process, future research might aim to bridge between tools by innovating on new formats that can be more easily shared between tools. For example, a recent application facilitates moving between Illustrator and D3, recognizing the iterative way in which these two different tools are often used to create a visualization-based story (Bigelow et al. 2017). Other tools like Lyra (Satyanarayan and Heer 2014b) aim to offer nonprogrammers the ability to create the same rich visualizations that often require the flexibility of visualization-based toolkits like D3 in an environment that also allows manipulation and customization as in Illustrator. Other opportunities include automated generation of intermediate frames to help smooth large jumps between visualizations in a linear story format, like a slideshow or animation, along the lines of Amini et al. (2017). Both research and industry development have already explored ways to reduce certain manual interventions in the telling a story phase by integrating automated functions for visualization annotation given a data set and corpus of relevant documents (Hullman et al. 2013, Gao et al. 2014, NarrativeScience). Future work might focus on integrating forms of story-focused automated suggestions into otherwise manually driven tools in order to reduce the time required for painstaking jobs like annotation formatting and placement.

**Collaboration support** is often critical across all phases of the storytelling process. Tight integration of collaboration and version management would enhance utility and lead to a much more effective and efficient process. While the ability to add comments to and share workbooks and other representations during analysis is built into systems like Excel and Tableau, the support for collaboration often fails to address all phases of the storytelling cycle and the multitude of tools involved. For example, how can a graphical user interface (GUI) data wrangling tool, or even a programming language toolkit, be extended to make it easier for collaborators to understand and reproduce the set of transformations that a colleague completed in the data cleaning phase? The development of more systematic processes and tool support for assessing data integrity could allow collaborators to specialize according to their interests and skills, so that the responsibility for the soundness of the data is shared by a team. Not unlike the challenges faced by a single designer working across tools with different formats, a team of collaborators may wish to divide work and work synchronously in different tools. The opportunity exists to envision tools that can facilitate the merging of assets in differing formats such as code, data, images, and text, for previewing the story as it emerges. A related challenge concerns supporting collaborators so they can easily provide feedback on emerging story materials created by others asynchronously and without requiring the expertise in the tool where the materials were created. Finally, most stories are considered finished once they are published, yet collaboration could continue in the form of audience feedback and subsequent design iteration. The audience for a story, such as an interactive graphic published online, may have additional design suggestions or analysis feedback worth incorporating into iterations (Hullman et al. 2015). Identifying ways to harness the audience's insight in common publishing venues (social media, news media, etc.) and to ease the efforts required to revise and republish designs are important avenues for future research.

## CONCLUSION

While visual data stories are maturing as a form of communicative visualization, there is a need to better understand the lifecycle of a story and support the practices and tools used for creating one. Towards this goal we have extended Lee et al.'s (2015) model of the iterative storytelling process and conducted semistructured interviews with nine professionals working in the area of visual data stories. Their rich accounts of particular experiences formed the basis for characterizing the major storytelling phases as exploring data, making a story, and telling a story. We then distilled the main roles involved in this process (data collector, analyst, scripter, presenter, and director) and described external factors that can impact it (audience, setting, medium, time and cost, and legal and ethical implications). Based on the interviews, we attempted to outline an overview of the systems, tools, and technologies that are being used in conceiving and constructing visual data stories. While this overview is not meant to be complete, it clearly shows that the tools and technologies being used can be highly idiosyncratic and often require specialized skills in software development and data mining. A mix of manual and automatic steps in data processing and graphics generation makes it difficult to support rapid story creation. The multiple skill sets needed, limited support for collaboration in creating a story, and a loosely joined toolchain can slow down the production of visual data stories and often prevent iterative refinement. Despite these challenges, there is a growing interest among authors and readers in data-based storytelling. The directions for future research and design include helping nonprogrammers to create stories with data, easing the process of finding relevant data, supporting the evaluation of stories, and enabling teams to collaborate internally as well as with their audiences.

## REFERENCES

Amini, F., Henry Riche, N., Lee, B., Hurter, C., and Irani, P. 2015. Understanding data videos: Looking at narrative visualization through the cinematography lens. *Proceedings of the ACM Conference on Human Factors in Computing Systems*, Seoul, Republic of Korea, pp. 1459–1468.

Amini, F., Henry Riche, N., Lee, B., Monroy-Hernandez, A., and Irani, P. 2017. Authoring data-driven videos with DataClips. *IEEE Transactions on Visualization and Computer Graphics* 23(1): 501–510.

Bigelow, A., Drucker, S., Fisher, D., and Meyer, M. 2017. Iterating between tools to create and edit visualizations. *IEEE Transactions on Visualization and Computer Graphics* 23(1): 481–490.

Boyatzis, R. E. 1998. *Transforming qualitative information: Thematic analysis and code development*. Sage Publications, Inc., Thousand Oaks, CA.

Burkle, L. A., Marlin, J. C., and Knight, T. M. 2013. Plant-pollinator interactions over 120 years: Loss of species, co-occurrence, and function. *Science* 339 (6127): 1611–1615.

Gao, T., Hullman, J. R., Adar, E., Hecht, B., and Diakopoulos, N. 2014. NewsViews: An automated pipeline for creating custom geovisualizations for news. *Proceedings of the ACM Conference on Human Factors in Computing Systems,* Seoul, Republic of Korea, pp. 3005–3014.

Gratzl, S., Lex, A., Gehlenborg, N., Cosgrove, N., and Streit, M. 2016. From visual exploration to storytelling and back again. *Computer Graphics Forum (EuroVis '16)* 35(3): 491–500.

Hearst, M., Laskowski, P., and Silva, L. 2016. Evaluating information visualization via the interplay of heuristic evaluation and question-based scoring. *Proceedings of the ACM Conference on Human Factors in Computing Systems*, pp. 5028–5033.

Hullman, J., and Diakopoulos, N. 2011. Visualization rhetoric: Framing effects in narrative visualization. *IEEE Transactions on Visualization and Computer Graphics* 17(12): 2231–2240.

Hullman, J., Diakopoulos, N., and Adar, E. 2013a. Contextifier: Automatic generation of annotated stock visualizations. *Proceedings of the ACM Conference on Human Factors in Computing Systems*, Seoul, Republic of Korea, pp. 2707–2716.

Hullman, J., Drucker, S., Riche, N. H., Lee, B., Fisher, D., and Adar, E. 2013b. A deeper understanding of sequence in narrative visualization. *IEEE Transactions on Visualization and Computer Graphics* 19(12): 2406–2415.

Hullman, J., Diakopoulos, N., Momeni, E., and Adar, E. 2015. Content, context, and critique: Commenting on a data visualization blog. *Proceedings of the ACM Conference on Computer Supported Cooperative Work*, pp. 1170–1175.

Jabr, F. 2013. Where the wild bees are: Documenting a loss of native bee species between the 1800s and 2010s. *Scientific American*, December 1, 2013.

Kandel, S., Paepcke, A., Hellerstein, J., and Heer, J. 2011. Wrangler: Interactive visual specification of data transformation scripts. *Proceedings of the ACM Conference on Human Factors in Computing Systems*, Seoul, Republic of Korea, pp. 3363–3372.

Kosara, R., and Mackinlay, J. 2013. Storytelling: The next step for visualization. *IEEE Computer* 5: 44–50.

Knudsen, S., and Carpendale, S. 2016. View relations: An exploratory study on between-view meta-visualizations. *Proceedings of the 9th Nordic Conference on Human-Computer Interaction,* Seoul, Republic of Korea, p. 15.

Lee, B., Riche, N. H., Isenberg, P., and Carpendale, S. 2015. More than telling a story: A closer look at the process of transforming data into visually shared stories. *IEEE Computer Graphics and Applications* 35(5): 84–90.

Luther, K., Pavel, A., Wu, W., Tolentino, J. L., Agrawala, M., Hartmann, B., and Dow, S. P. 2014. CrowdCrit: Crowdsourcing and aggregating visual design critique. *Companion Proceedings of the ACM Conference on Computer Supported Cooperative Work & Social Computing*, Seoul, Republic of Korea, pp. 21–24.

Qu, Z., and Hullman, J. 2016. Evaluating visualization sets: Trade-offs between local effectiveness and global consistency. *Proceedings of Beyond Time and Errors: Novel Evaluation Methods for Visualization*, Seoul, Republic of Korea, pp. 44–52.

Roberts, J. C., Headleand, C., and Ritsos, P. D. 2016. Sketching designs using the Five Design-Sheet methodology. *IEEE Transactions on Visualization and Computer Graphics* 22(1): 419–428.

Satyanarayan, A., and Heer, J. 2014a. Authoring narrative visualizations with ellipsis. *Computer Graphics Forum* 33(3): 361–370.

Satyanarayan, A., and Heer, J. 2014b. Lyra: An interactive visualization design environment. *Computer Graphics Forum* 33(3): 351–360.

Segel, E., and Heer, J. 2010. Narrative visualization: Telling stories with data. *IEEE Transactions on Visualization and Computer Graphics* 16(6): 1139–1148.

Stolper, C. D., Lee, B., Riche, N. H., and Stasko, J. 2016. Emerging and recurring data-driven storytelling techniques: Analysis of a curated collection of recent stories. Microsoft Research Technical Report MSR-TR-2016-14.

Trielli, D., Mussenden, S., Stark, J., and Diakopoulos, N. 2016. Googling politics: How the Google issue guide on candidates is biased. *Slate*, June 7, 2016.

# CHAPTER 8

# Organizing the Work of Data-Driven Visual Storytelling

Christina Elmer

*Spiegel Online*

Jonathan Schwabish

*Urban Institute*

Benjamin Wiederkehr

*Interactive Things*

CONTENTS

## INTRODUCTION

Over the last 10 years, we have seen a remarkable shift in the use of data and statistics. Data is becoming easier to collect and publish, and has quickly become common currency for businesses, governments, and other sectors to conduct and improve their work. The importance of data visualization has quickly followed the increasing value and availability of data, and the evolution to faster and cheaper computer speed and memory.

Visualizing data can take many forms, from traditional static graphs to online data tools where users can explore and download data for their own needs. In between are narrative data stories that combine text and data that are in many ways more complex and sophisticated than in the past.

As the demand for better data visualizations and these narrative stories increases, how should groups and firms organize their workflow processes? What skill sets should they look for in their employees? What structural changes should they implement to ensure that data are making their way through the collection, analytics, and communication phases? What productivity, communication, and production tools are organizations using to visualize their data and to tell stories with data?

In this chapter, we report results from interviewing more than a dozen people across three main fields: design firms, media, and nonprofits and nongovernmental organizations. (A copy of the questionnaire we used can be found in the Appendix at the end of this chapter. For some, we also followed-up with additional interviews.) Design studios help organizations present and explain complex topics to a wide range of audiences. They tend to be commissioned to do work on behalf of their clients, but often leave the in-depth data analysis work to the client. Media organizations use data-driven visualizations to illustrate their findings in a variety of ways across multiple platforms (e.g., print, desktop, and mobile devices). We interviewed journalists in small, independent teams and larger newsrooms that have a more integrated model. Finally, we interviewed organizations in the nonprofit, nongovernmental, and local government areas. These organizations are producing a variety of products and information, often in a variety of forms, but their business and production models differ from the other two groups. Overall, we seek to understand how these groups manage their workforce, their data, and their processes to communicate their work (and the work of their clients and partners) in the best way possible (a discussion about tools and the lifecycle of a data project can be found in Chapter 7 entitled "From Analysis to Communication: Supporting the Lifecycle of a Story").

Across these different sectors and firm structures, we find striking similarities in staffing and the flow of work through the organization. Three takeaways seem keenly important. First, there is no "unicorn" employee that can do or handle all the work for data-driven projects (though one of our media organizations strives to train their journalists as unicorns). Project production requires a variety of skills in data, design, Web development, programming, and more, and all of our interviewees mentioned that it is rare that a single person would have all of those skills or be able to perform all those necessary functions for all of their projects. Second, there is no single tool or even agreed-upon suite of tools to best facilitate communication and organization. The groups we interviewed used a mix of data, chat, and organization tools, but they did not agree on a single suite. Third, each group is in a constant state of development, change, and learning. These groups have each acknowledged that they are still learning best practices and processes, and adapting to changing landscapes in their particular sectors.

We also found a variety of differences across the three areas. First, the three groups are centered around different types of people. The design studio tends to be based on developers and designers, media in journalism, and nonprofit/nongovernment organizations (NGOs) are based around analysts and researchers. Each organization reports they need each of these other skill sets in their workflow, but the core skill upon which they are built—and hence the culture that permeates each organization—is fundamentally different. Second, the business and funding models of each differ in a variety of ways. Design studios and nonprofit/NGOs are often seeking funding and grants, while (most) media organizations are relying on advertising and other forms of revenue. We did not explore these differences at length in our interviews, but suspect those forms also drive differences in project management and organizational structure.

Overall, telling data-driven stories remains a function of an organization's mission and audience. Whether an organization is rooted in design, telling stories, or data analysis, each needs each piece to effectively communicate their work and analysis to their audience. In this chapter, we seek to explore the similarities and differences between these disparate groups and to offer insight into what makes those efforts successful.

## DESIGN STUDIOS

Over the past decade, we have observed the rise of a new kind of design studio that applies the theories and practices of the field of information visualization to the space of consumer products and services. These agencies help organizations from both the public and private sectors present and explain complex topics to a wide range of audiences. With a multidisciplinary team composition, they combine skills from information visualization, interaction design, interface design, Web development, software engineering, and data science. The balance between design-oriented and technology-oriented capabilities will influence the nature of the work a team produces—what type of problems it is able to solve and what type of solutions it is equipped to deliver. Such diverse groups of practitioners are able to serve the audience's expectations and needs with the right balance between simplicity and sophistication.

These new design studios are responding to an ever-increasing need to understand, evaluate, and communicate more and more data. We posit that there are four primary factors driving changes in this sector. First, the relative complexity of creating data-driven stories that are accessible, engaging, and compatible across all device categories (e.g., desktop, tablet, mobile, and virtual reality (VR)) demands well-trained and experienced designers and developers. Second, finding professionals who possess all of the required skills is extremely rare. High-quality work is usually and ideally done in collaboration between specialized experts. Third, hiring a full-fledged creative team requires significant financial investments, which might not be feasible for all organizations. Finally, establishing a fluid process across the entire lifecycle of a story—from ideation to content creation to designing the appropriate format for developing and deploying the final product—may be a fundamental change to the structure of the organization. Design studios, therefore, can help organizations to overcome—or rather bypass—these challenges on an on-demand basis. The level of engagement between the client and the vendor may differ from project to project. The typical scope of work for the studio is limited to the design and development of visual and interactive artifacts, while the client provides the data together with initial insights. But this separation of work does not exclude the design studios from engaging deeply with the data and supporting analysis and interpretation. Successful studios are those that understand how to translate data into compelling stories and how to do it in collaboration with potentially large and often slow-moving organizations.

## Organizational Structure

We interviewed professionals from six different design studios working across sectors and regions (the list of organizations can be found in the Appendix at the end of this chapter). All six of the interviewees are founding partners of their studios with a clear understanding of the business side of their operations. Three of our interviewees hold the position of "creative director" and are deeply involved in the creative aspects within their teams. The other three act as "managing directors" or "chief executive officers" and oversee the whole studio operation.

All interviewees have indicated that they intentionally structure their teams in flat hierarchies with a cross-disciplinary spectrum of professionals. The studios are led and managed by technically skilled personnel. The professionals we interviewed do not have backgrounds in economics, finance, legal, or business administration, but instead have technical and design expertise.

We have observed a distinct separation between design-oriented and technology-oriented practitioners in the design studios we interviewed. Teams are usually constructed with a lead designer on the one side and a lead engineer on the other. We did not find any reference to individual team members that combine both competencies in a leadership capacity; in other words, we did not find the often sought after "unicorn" in these six studios, a single person capable of designing and developing sophisticated visualization or storytelling experiences. Instead, the collaborative nature of the type of work was reflected in most of our conversations. In addition to the design and technology leads, these teams also consist of more specialized personnel including data analysts, copywriters, and illustrators, as well as more general personnel like project managers.

When it comes to technical work on a specific project, each studio reported a close collaboration between the content, design, and technology teams. That said, the driving force in the creative process might differ. For Accurat, the design team assumes this role: "The design team I direct is at the core of our organization. The design process informs both the data analysis and the development process," says Giorgia Lupi, the agency's design director. Whereas for Column Five, the most complex data storytelling is handled by the interactive department where the development team is capable of working with large, messy data sets.

## Skill Sets

We found a few skills that are core to each team, as well as a list of additional secondary skills. The core of all teams that we interviewed involves interaction designers, visual designers, and front-end developers. This constellation of experts seems to be well-suited and sufficient to build best in class experiences for the Web. As expected, we also found complementary skills such as content production, e.g., photography or videography, copyediting, back-end development, and data analysis. Although secondary, these skills can add a lot of value to data-driven stories. For example, a visualization driven by a database interface might provide a more comprehensive and personalized experience over an aggregated visualization, or a series of carefully illustrated animations will excite interest as opposed to a set of dull, static graphics. That being said, we found little mention of specific team members dedicated to social media or general digital services or management such as website development and maintenance. The leaders of these design studios seem to be confronted with the challenge of building a team that is rich with a variety of skills and efficient and lean at the same time.

Not every studio housed all of the needed skill sets. We found a significant use of freelance experts to help solve problems and complete projects. For design-oriented teams, this could mean outsourcing the development of heavy-duty applications; for others, this could mean hiring copyeditors or marketing professionals. Interestingly, most of the studios we interviewed did not have an in-house team of people to conduct subject or background research, collect data, create multimedia content, or managers to help shepherd the work through the organization from inception to publication. These tasks are primarily conducted by the project teams within the client organizations or in collaboration with additional third-party agencies or studios.

## Tools and Technologies

The design studios we interviewed apply a broad spectrum of tools to produce and coordinate their work. Across our selection of companies, the overlap between the most popular tools is significant with only a few exceptions of tools only used by individuals. (We provide a list of mentioned tools and technologies, a short description, and URL for each in the Appendix at the end of this chapter.) Even though most of these teams work in a shared office space, the people they work for or collaborate with might not always be accessible face-to-face. Digital tools to communicate and to coordinate their work are indispensable. Colocation seems to have little impact on the diversity of the amount or types of tools used by these teams. Teams rely on different tools to support the individual steps of their workflow: to share files, coordinate tasks, track time, schedule meetings, create static design mockups, prototype interactive interfaces, and implement functional applications. Here we consider the current stable of tools in use while recognizing that these can change quite rapidly and that organizations need to continually evaluate new tools that can make the work more productive.

To store and share files, most teams rely on either Google Drive or Dropbox. To communicate with each other, the most used channels are email, Skype, or Slack. To manage and coordinate open tasks, some of the studios use dedicated project management applications like Basecamp, Trello, or TeamworkPM, while others rely on physical Kanban boards in their office space. For time tracking, which is essential for project-based contracts, the two most-often used applications are Harvest and Toggl. For scheduling activities and milestones the most popular solutions seem to be Google Calendar and Forecast. Both applications allow team leaders to get an understanding of the future workload and to coordinate the assignments for each team member.

The most widely used tools among the design departments in the design studios we interviewed are the Adobe Creative Suite including Photoshop and Illustrator as well as Sketch by Bohemian Coding. For prototyping interactive visualizations and interfaces, there is currently a broad selection of tools to pick from: InVision, Principle, and Framer seem to be the most popular ones.

For developing the front-end of data-driven applications for the Web, most teams use HTML, CSS, and JavaScript with the support of open-source libraries and frameworks. D3.js was mentioned as one of the most important components for interactive data visualizations while Angular and React were mentioned as essential to structure more complex applications. The specific configuration and structure of the code base depends on the coding principles and styles of the different team members.

## Process and Project Selection

In comparison to the other two groups of organizations that we interviewed, design studios primarily work for clients on a per-project basis and they therefore have to rely on inquiries and commissions from clients. Typically, these inquiries come in two different forms: either a direct inquiry from a client who sought out a specific agency to collaborate with, or a request for proposals (RFP), where different studios may apply to win the contract. Among the studios that we interviewed, the first approach is preferred as it involves less guesswork and competitiveness when it comes to the financial offering.

Most of the studios cultivate a selection process during which they evaluate how well a project might fit their team. Mentioned factors include the impact a project has on the client or the end-user, the alignment of the project's vision with the studio's strategic mission, the fit between the project's requirements and the team's skills, and the personal compatibility between the client and the studio's teams.

The work within a single project is described by most interviewees as highly interdisciplinary and collaborative. This not only includes professionals from different areas within the studio but also different departments and roles from within the client. To coordinate the work, the studios usually rely on a dedicated project manager within their team with a corresponding project management role within the partnering organization.

Typical time frames for projects that are being taken on by the studios we interviewed range from 1–3 weeks all the way up to 3–6 months. We identified a clear correlation between preferred duration and team size: generally, larger studios prefer to work on more long-term projects while smaller ones are more comfortable and used to work on fairly short-term assignments. We expect one reason for this to be the overhead necessary to onboard a bigger team onto a project that may be infeasible for very short assignments.

## Reflections and Lessons Learned

The work in an agency model poses a mix of organizational, technical, and temporal challenges. The leaders of these design studios have to succeed in the technical work as well as managing the success of the business itself.

This requires the individuals to switch their perspective from design and development to finance and legal, and to mediate between these sometimes competing areas.

Based on this observation, one of the two hardest challenges is running successful projects alongside running a successful business. Although a project success contributes to the bottom line of the overall business, the health of the company needs more than solid published work. Building and cultivating a team needs time, as do business development, customer acquisition, and administrative tasks.

The second big challenge is managing the collaboration between the studio's team and the client's team. Oftentimes these two teams have competing objectives: the client is focused on getting the most out of the collaboration while the design studio strives to optimize its resources to serve not only one, but a multitude of clients at the same time. As a result, changes to the scope or the timeline during the project are a recurring point of friction. The most often given answer to the question on how to resolve this issue is that everything becomes easier to manage with clear, transparent, and timely communication, while giving everybody the opportunity to understand what is going on in a project and to make smart decisions while managing the work.

When we asked the leaders of design studios about their recommendations for doing great work in this industry, we heard inspiring advice across a broad spectrum. They ask us to embrace complexity and remind us that data-driven storytelling is about making complexity accessible, not about simplifying it. They challenge us to learn about the history, methodology, and theory, but then also encourage us to consciously break away from standardized visual models and preexisting beliefs in order to produce something ambitious and compelling for our audience. Finally, they invite everyone to engage in the community and be part of the conversation in order to learn and contribute.

## MEDIA ORGANIZATIONS

The second group we interviewed consists of media organizations. Media organizations use data-driven visualizations to illustrate their findings in both static and interactive ways. In a media landscape that is rapidly changing, these formats help feed different distribution channels and to express an organization's approach to produce exclusive output. Many media outlets have set up specialized teams to produce visual stories based on their data-driven investigations.

We interviewed six professionals working within or closely related to media organizations. Three described themselves as data team leaders, and one each as an interactive team leader, editor of visuals, and data journalist. These different job titles reflect the different strategies media outlets now follow to implement visualizations with a news story and within the organizational structure of the newsroom. The groups we interviewed ranged from small, independent teams to more integrated models where one large team unites all the visually driven editors, designers, and programmers within the newsroom (the list of organizations can be found in the Appendix at the end of this chapter).

These organizations also reflect different publishing strategies, ranging from an explicit focus on online publishing to a mixture of different channels, be it a printed newspaper or digital platforms. Our interviews, therefore, shed light on a fully differentiated media sector without claiming to capture every possible strategy. But even among the evaluated and very diverse teams, shared characteristics and challenges can be clearly identified.

## Structure and Organization

Three of the media organizations we interviewed are daily newspapers currently publishing much of their content online. We also interviewed one television organization which also publishes content online. Two organizations are nonprofit newsrooms, who investigate and publish topics by partnering with other media outlets that tend to be more traditional. These nonprofit media organizations have different publishing requirements than the traditional media groups such as using specific tools (sometimes open source) or other collaboration requirements.

The reach of each organization's data-driven stories differs widely. Some focus on a single city whereas others publish topics of great interest for entire countries. These geographic restrictions are becoming less and less important as news articles are now read and easily shared online. We also found that these teams know each other quite well and often share strategies and lessons learned.

Within their organizations, the six teams are placed quite differently within the larger organization. Most of them work fully integrated into the overall newsroom and its routines. At the same time, they are often given freedom to conduct investigations that need more time and resources relative to breaking news or shorter news pieces. In one case, this is reflected by a comparatively clear segregation and a rather disruptive approach.

This particular team is not integrated into the news cycle at all and works on self-chosen projects in an independent way. Of course, editors are invited to join in for single pieces, but a closer consolidation is obviously avoided to protect the team's innovative strength.

The implemented hierarchies of the teams we interviewed depend mostly on their size. The already mentioned segregated team, for example, works without any kind of hierarchical structure, whereas one larger team of about 50 people is broken into subteams and has editors managing projects that cross news desks. Between those extremes, a common team structure we observed in media organizations consists of a team lead with a journalism-shaped background and a second hierarchical level including developers, designers, and other journalists. This tendency towards flat hierarchies with only two levels seems to reflect both the rapid development of this new newsroom space and the need to include and manage a broad range of competencies.

## Skill Sets

Journalism, design, and code seem to be the core skills required for realizing data-driven visualizations within the media organizations we interviewed. In this regard, design can include both visualization design and interaction design. Code refers to developing interactive visuals as well as their underlying databases. And the journalism part includes the selection of topics, further investigations, and the content-related composition of the final publication. In addition to that, both code and journalism skills should include data analysis competences. Some experts described journalism, design, and code as equally important, whereas Scott Klein (from *ProPublica*) declared journalism to be the make-or-break skill: "We can get the other two wrong and recover, but if we get the facts wrong, the other two don't matter."

Each of the organizations emphasized the importance of efficient communication, especially because team members are inherently interdisciplinary. However, one interviewee noted the lack of a project manager; another explicitly noted that the team consisted of coding and user experience expertise, but less background in traditional journalism; and yet another interviewee remarked that overall organizational skills were absent from the team.

In contrast to the interdisciplinary approach, *ProPublica* follows a different strategy. Within their "news applications desk," developers are "responsible for an entire project, from inception through publication.

They need to build the server code to support their ultimate vision without slow negotiation or the friction of brain-to-brain communication, and they need to design the presentation because they're the person who understands the material and the visual story possibilities best." In this case, the team manager recommended searching for people with two of the necessary skill sets and to teach him or her the third.

## Tools and Technologies

For communicating with colleagues and project partners, most of the teams use Slack or the open source, self-hosted alternative Mattermost. Google Hangouts, Google Drive, and GitHub were mentioned multiple times, as well as the encrypted chat system Jabber OTR and encrypted emails. These safety arrangements seem to be most important for the teams who often work together with investigative reporters and confidential sources.

Two of our interviewees noted that these communication tools can never be a substitute for the direct communication within a team room. The team led by Scott Klein even has a rule that no digital task can be a surprise: "Everything starts as a real-world conversation, and digital project management tools are simply a way of tracking what everybody's doing."

For managing tasks and projects, half of the teams work with Trello, a Web-based application that has been developed based on the Kanban management system. The latter can of course also be implemented with the help of other tools like a whiteboard or Wekan, an open-source Kanban software. Throughout the six teams in our sample, one of these alternatives can be found in every questionnaire.

Given the many different tools and platforms teams are using for communication and management tasks, one need seems to be a better integration across the entire workflow. For example, Ulrike Köppen, from Bavarian Broadcasting, pointed out that an "umbrella tool for project management and the organization of the material" would be extremely helpful.

## Process and Project Selection

Being part of a larger newsroom, all six teams are integrated into the decision-making processes about the topics they are investigating. Ideas emerge within the team or are suggested by reporters and editors from other departments. Most of the experts we interviewed described both ways as important for their workflows. Thus, a certain openness to pitches from outside the team seems to be crucial for a successful integration.

The decision for or against such an idea is obviously made at different hierarchical levels. The more independent teams mostly make their own decisions, whereas other teams are dependent on instructions from a manager or editor. The latter decision is very much linked to central journalistic relevance criteria as well as to the overall strategy of the media organization. One of the interviewees from a nonprofit newsroom noted that the most important consideration is the impact of the reporting.

Within single projects, the teams we interviewed usually collaborate with other departments. As both a requirement and a consequence, most experts emphasized the value of close relations to other units within their organization, depending on their distribution channels. Thus, for the broadcast data team, radio and television departments are important partners, whereas others join forces especially with design or video experts.

The typical time frames of producing a data-driven visualized story range from 2 weeks up to several months. According to one interviewee, single projects can be done by his team in 1 week whereas others need 18 months to be finished. However, the mean duration for this kind of project is typically about 3 months. In our sample, the longest time frames were reported by nonprofit newsrooms who work independently from a day-to-day news cycle.

## Reflections and Lessons Learned

Within a news-orientated environment, long-term project planning necessary for data-driven investigations can pose a huge challenge. Several experts described regular conflicts between the existing commitments, new project ideas, and real-world events. Priorities have to be adjusted constantly, also because of the multiple angles from which ideas can reach the team.

The second crucial challenge seems to be the overall communication of the team's projects, skills, and outputs within the organization. Even Scott Klein at *ProPublica* emphasized this task: "We are working to get better at communicating our priorities and explaining our work to the wider newsroom." In smaller teams, this task is often taken over by the journalist who acts as a project manager and simultaneously works on the content.

When asked about their strongest recommendations, many of the interviewed experts described two lessons. First, accept a constant change of the ecosystem in which the team is working. By being open to learning new processes and products, change can be made more smoothly and utilize skills from the entire team. This openness also enables individuals to share their work and experiences. As one key strategy, *The Guardian's* Xaquín González Veira also recommended to always "put yourself in the shoes of your audience."

The second lesson is the importance of teamwork and its composition. Nearly all of the respondents—from both small and large newsrooms—argued in support of diverse, interdisciplinary teams. Data team leader Ulrike Köppen noted that, "The more varied the input, the better!"

Finally, in addition to reasonable recommendations like "Go viral. Get covered. Win awards."*, one special tip may be well qualified to conclude this section. In a media ecosystem of rising economic pressure and shrinking profits, Gonzalez pointed at the general mindset creative teams need to produce brilliant products: "Have fun and take your work with a pinch of humor—take your job seriously, but not so much yourself."

## NGOs AND NONPROFITS

The third set of organizations we interviewed were groups in the non-profit, NGO, and local government spaces. In some ways, these groups are similar to media organizations in that their business model involves producing a variety of products and information, often in a variety of forms.

These groups also differ from media organizations in three significant ways. First, they are often seeking outside funding and sometimes responding to funders or grantor requests. This means that they may be writing for a more-specific audience or an audience with particular expertise. They may not always get to choose the particular projects they wish, nor are they always able to choose the format in which those projects are published.

Second, staffs at these organizations may, overall, have higher levels of statistical or mathematical expertise than the design studios or media groups we interviewed. The primary goal of these organizations, after all,

---

* Want to start a small data journalism team in your newsroom? Here are 8 steps, Scott Klein, 7 April 2016 (accessed 15 June 2016). http://www.niemanlab.org/2016/04/want-to-start-a-small-data-journalism-team-in-your-newsroom-here-are-8-steps/.

is to evaluate and assess policies or programs in a (primarily) quantitative way. Thus, such organizations often consist of staffs with advanced degrees in research and quantitative fields such as economics, sociology, demography, and political science. On the one hand, these organizations have groups of people able to conduct and produce content. On the other hand, however, those same staff members typically do not have the design or communications background necessary to effectively communicate their work. By contrast, and as we noted earlier, the design studios we interviewed did not typically have personnel to research or collect data.

Third, the mix of research/scientific staffs and communication/ outreach/media staffs can also result in stresses and conflict that may not necessarily exist in sectors or organizations where the staff is more homogeneous. While researchers may emphasize the analytics, nuance, and details of their data analysis, communication and design staffs may emphasize the need to communicate that work to an audience or to several audiences; those two goals may not always seamlessly mix.

## Structure and Organization

We interviewed five organizations for this section: three are nonprofit research institutions that focus on a wide array of issues including health, immigration, and economic policy—Mathematica Policy Research, The Pew Charitable Trusts, and the Urban Institute (disclaimer: one of us, Jonathan Schwabish, works at the Urban Institute); the Office of the Deputy Mayor for Planning and Economic Development (DMPED) in Washington, D.C., which helps the mayor carry out his or her economic development plan; and the Inter-American Development Bank (IDB), an international organization that helps provide funds and support to countries in Latin America. Two of the five people we interviewed have "vice president" in their job titles, and two others are "program analysts" or "officers."

The five organizations range in size from about 80 people to more than 3,000 people. The communication teams also vary in size and range from about 6 people to more than 60 people. The makeup of those teams, in terms of skill sets and specialties also differ. For two of the nonprofit organizations, we interviewed representatives from the central communications departments. Those departments consist of a wide range of members and include a variety of skill sets such as design, digital communications, editorial, and external affairs. We interviewed an officer with The Pew Charitable Trusts who works on a specific project within the larger organization. That group consists of seven people divided into a research side and a communication side; all seven team members, however, work together "in developing the research agenda and strategic direction of the project."

The communication and outreach department at the IDB focuses slightly more on supporting the work of the research staff instead of publishing content outside the organization; instead, they focus on "coordinat[ing] training and techniques for making effective presentations, including storytelling and script-writing."

## Skill Sets

Each organization specifically mentioned some staffing challenges. Researchers at these organizations typically use statistical packages such as SAS, SPSS, Stata, and R, but have less expertise in tools and coding languages used to produce online content such as HTML, CSS, and JavaScript. Some of the groups mentioned bottleneck issues because of these different skill sets and were constrained in their hiring decisions because of funding or other issues. One interviewee mentioned that she had learned Python, Web development, d3, and JavaScript on her own, and while her superiors were supportive of her efforts, no external professional development training was offered.

For some of the larger communications departments, we observed a range of other skills needed to not only help produce the final product, but to help shepherd the work through from conception to delivery. Those organizations include people who had expertise in writing and editing, social media, digital services, and management.

## Tools and Technologies

Each organization listed a variety of tools they use to create and organize their workflow, as well as supporting collaboration both within and across departments.

For content creation, the usual set of tools were listed: d3 and JavaScript, R, Tableau, ArcGIS, and Stata, as well as standard productivity tools from Google and Microsoft. With respect to collaboration and organization, each organization is using a suite of tools.

- Asana, HipChat, Jabber, and Slack are being used for internal chat.

- All of the organizations mentioned some kind of cloud-based storage tool, but Box was the only one named.

- A variety of other customer relationship management (CRM) and organization tools were mentioned and include Evernote, Jira, Salesforce, SmartSheet, and Work Front.

Nearly all of the organizations mentioned some kind of short "stand-up" or "daily scrum" meeting where parts or the entire team gathered. Two of the organizations mentioned using a whiteboard to help share and categories ideas, projects, and responsibilities. One organization mentioned some internal tools, but did not elaborate.

None of the people we interviewed seemed wholly satisfied with their suite of tools. Project workflow and organization are clearly challenges, especially for groups that are moving work between divisions or between roles within a single division. One interviewee responded that, "tools are only as effective as the user input that is provided, and we still struggle to ensure these systems are regularly updated. No one system can provide all the necessary collaboration, sharing and tracking, and we struggle to maintain all these systems at once."

## Process and Project Selection

Each organization publishes an array of product types that range from research reports, briefs, data visualizations, blogs, maps, graphs, and multimedia stories. Some of those reports are externally facing, but some are for internal use or are confidential for clients or partners. Projects are undertaken for a variety of reasons: client need and consultation, funder or organizational priority, and "newsworthiness." Projects range from hours to months. One organization noted that they prioritize their projects "based on their urgency and our skill sets, and we keep a backlog list. In this way we operate a bit like an internal data consulting shop."

DMPED, the Washington, D.C. agency charged with economic development, is also responsible for building and maintaining an open-source data dashboard. That dashboard is regularly updated and the team has constructed a workflow that enables them to quickly and easily update the dashboard as new data become available (more on the tools they use below). Perhaps unlike the other groups, DMPED responds to both client requests (other government agencies and the mayor's office) and requests from the public.

### Reflections and Lessons Learned

We asked each organization for their strongest recommendations to newcomers to the field of data-driven storytelling and communication. Respondents provided us with a variety of thoughts, which we have categorized into three broad buckets.

First, communications should not be thought of as a secondary part of the process of communicating research: "strategic support is essential" as one person put it. Respondents recognized that there often exists tension between the scientist/researcher side of the organization and the communications/marketing/design side of the organization. Those two groups have very different skill sets and the way they think about presenting their work to their audiences. One respondent remarked that they are often "stuck in one part of the research-policy-practice loop, but it's only through connecting the entire loop that everyone's products can be amplified."

Second, embrace new tools and technology and let people play with them. As one respondent noted, "Evolve constantly by allowing others to join you. Every year is different from the one before—new products, new people, new approaches." Data visualization tools like Excel, Tableau, and d3 may not be accessible for everyone in the organization, but tools that can be used by people who may not have a coding background can be used as a "playground" for innovation and change.

Finally, do not get frustrated; have patience, and build understanding. These and similar phrases popped up throughout the written and oral interviews. Respondents were eager to emphasize that the various skills and cultures across divisions in an organization can cause headaches and fights, but by having patience and explaining how one group can help another and vice versa, the product improves. As one person said: "Be open, take feedback, and refine with each new project."

## CONCLUSION

The organizations we examined here show both similar and specific features when it comes to producing data-driven visuals. Similarities arise mainly from the special requirements going along with designing, developing, and publishing such innovative formats. But depending on the organizational environment, this challenge is approached rather differently.

The more complex the overall scope of the organization, the more special skills and roles seem to be needed within these teams.

Design firms and professionals in this field can focus on creating the visualizations itself, omitting, for example, the publication strategy of the final product. This leads to teams with two clear priorities: design and technology. Both in media companies and nonprofits, teams tend to be more diverse, including also journalists, communication experts, or researchers. Nevertheless, there are two main structural similarities: all teams are composed cross-disciplinary and work within rather flat hierarchies, often consisting of only two levels.

The interdisciplinary composition of these groups results from their main priorities and fields of work. All practitioners emphasized the importance of core skills like data handling, visual design, and developing. Beyond that, the production of data-driven visuals requires additional skills like data research and collection, content creation, and process management. In this regard, design studios commonly collaborate closely with their clients who can contribute the expertise required.

On the other hand, media companies and nonprofits have to consider these skills already during the setup of their teams. Since both their size and budget are often limited, employees with cross-disciplinary interests and autodidact approaches are usually sought after. One team even tries to recruit solely potential unicorns, developers who are responsible for every working step of an entire project.

Whereas design studios are absolutely free to choose their projects, teams within media companies and nonprofit organizations also follow external suggestions. Nevertheless, the project selection criteria clearly overlap among the three groups. The potential impact of a data-driven visual story seems to be one crucial attribute, as well as its fit to the strategy of the organization. Furthermore, teams consider how well requirements and skill sets go together—a consideration most common in units that work independently, such as design studios or freelancers.

Project time frames differ strongly from hours up to several months. Asked about their mean time span, several media and design firms mentioned 3 months. Due to organizational reasons, larger teams within design studios tend to choose more long-term projects. Among media organizations, such larger topics are especially preferred by teams with a greater distance to the daily news cycle.

To support their workflows and production processes, the interviewed teams use a wide range of tools and technologies. They need platforms to share files and to communicate, to schedule meetings and to organize their workflow, to do interactive design and prototype applications. When collaborating with investigative reporters, teams additionally have to ensure a high level of confidentiality. Many of them reported difficulties in keeping all the systems maintained and updated, hoping for more integrative tools to be developed, ideally suitable for the entire workflow.

From an overall perspective, the professionals we interviewed described different sources of friction. Most of them seem to emerge at the interfaces to other units or partners that are working with different skills or objectives. To handle these challenges, many practitioners recommended a patient, continuous, and transparent communication, be it with project partners or other departments within their organization. Hence, teams should always be open to feedback and constantly explain their work to build understanding, foster collaborations, and improve their results. Despite such lessons learned, many interviewees expressed themselves as data-driven enthusiasts, recommending to be creative and ambitious, to engage in the community, and to share insights.

## APPENDICES

## GLOSSARY AND DEFINITIONS

**Interaction designer:** A professional who designs interactive digital products and services with a focus on the structure, function, and behavior of the user interface.

**Visual designer:** A professional who designs static or interactive products and services with a focus on the aesthetics of the graphical user interface.

**Front-end developer**: A professional who develops the user interface of Web applications and websites.

**Creative director**: A senior professional responsible for all creative operations of an organization to include staff supervision and work production.

**Data journalist:** A journalist who uses data sets for the investigation of a topic and for the visualization of his or her findings.

**Managing editor:** A senior editor who usually reports to the editor-in-chief and manages all production and publication stages.

## TOOLS

| Name | Description | Link |
| --- | --- | --- |
| **Chat Tools** | | |
| Google Hangouts | Communication platform developed by Google allowing users to video chat and use instant messaging | https://hangouts.google.com/ |
| HipChat | Group and private chat, file sharing, and integrations | https://www.hipchat.com/ |
| Jabber OTR | Off-the-record (OTR), encrypted communication with the instant messaging service Jabber, now XMPP | https://www.jabber.org/ http://xmpp.org/ |
| Mattermost | Open source, self-hosted alternative to Slack | https://www.mattermost.org/ |
| Slack | Team communication software with both open and private channels, cloud-based | https://slack.com/ |
| Skype | Application that provides video chat and voice call services | https://www.skype.com |
| **Productivity/Management Tools** | | |
| Asana | Web and mobile application for project management and collaboration | https://asana.com |
| Basecamp | Web and mobile application for project management and collaboration | https://basecamp.com/ |
| Evernote | Web, mobile, and desktop application designed for note taking, organizing, and archiving | https://evernote.com |
| GitHub | Web and desktop application for Git repository hosting and source code management | https://github.com/ |
| Harvest | Web, mobile, and desktop application for time tracking, invoicing, and reporting | https://www.getharvest.com/ |
| InVision | Web and mobile application for prototyping, collaboration, and design workflow management | https://www.invisionapp.com/ |
| Jira | Web and mobile application for bug reporting, issue tracking, and project management | https://www.atlassian.com/software/jira |
| Salesforce | Web and mobile application for case and task management including tracking and networking tools | https://www.salesforce.com/ |
| Smartsheet | Web and mobile application for project management including storage and collaboration | https://www.smartsheet.com/ |
| Toggl | Web, mobile, and desktop application for time tracking | https://toggl.com/ |
| Trello | Web and mobile application based on the Kanban management system | https://trello.com/ |

(*Continued*)

| Name | Description | Link |
|---|---|---|
| Wekan | Open-source application based on the Kanban management system, similar to Trello | https://wekan.io/ |
| WorkFront | Web application for work and project management including issue tracking, document management, time tracking, and portfolio management | https://www.workfront.com/ |
| **Statistics Programming Tools** | | |
| ArcGIS | Geographic information system (GIS) for working with maps and geographic information | https://www.arcgis.com/features/index.html |
| R | A language and environment for statistical computing and graphics | https://www.r-project.org |
| SAS | Software for advanced analytics, multivariate analyses, business intelligence, data management, and predictive analytics | http://www.sas.com/ |
| Stata | A general-purpose statistical analysis package for statistical analyses, data management, graphics, simulations, and custom programming | http://www.stata.com/ |
| Tableau | A business intelligence (BI) tool used to create reports, charts, graphs and dashboards | http://www.tableau.com/ |
| **Web Programming Tools** | | |
| Angular | A structural framework for dynamic Web apps that can be used to extend HTML's syntax | https://angularjs.org/ |
| CSS | Style sheet language used for describing the presentation of a document written in a markup language like HTML | http://www.w3schools.com/css/ |
| D3 | JavaScript library for producing dynamic, interactive data visualizations | https://d3js.org/ |
| HTML | Hypertext Markup Language, a standardized system for tagging text files to achieve font, color, graphic, and hyperlink effects on the Web | http://www.w3schools.com/html/ |
| JavaScript | A high-level programming language for HTML documents | http://www.w3schools.com/js/ |
| React | A declarative JavaScript library for building component-based user interfaces | http://reactjs.org |
| **Design Tools** | | |
| Adobe Creative Suite | A software suite of graphic design, video editing, and Web-development applications | http://www.adobe.com/ |
| Framer | A software to create animated or interactive prototypes with minimal programming knowledge | https://framerjs.com/ |

(*Continued*)

| Name | Description | Link |
|---|---|---|
| Principle | A software to create animated or interactive prototypes without programming knowledge | http://principleformac.com/ |
| Sketch | A software specialized for the design of user interfaces for websites and applications | https://www.sketchapp.com/ |
| **File Hosting Platforms** | | |
| Box | File hosting, synchronization, collaboration, and content management service | https://www.box.com |
| Dropbox | File hosting, synchronization, collaboration, and content management service | https://www.dropbox.com |
| Google Drive | File hosting, synchronization, collaboration, and content management service | https://drive.google.com/ |

## INTERVIEW QUESTIONNAIRE

*On Organizational Management for Data-Driven Storytelling*

1. About yourself
   What is your official job description?
   Please describe your job in one sentence.

2. About your organization
   Please describe the main goal(s) of your organization.
   How big is your organization?
   What is your organization's main product and audience?

3. About the capabilities in your team
   Please describe where within the entire organization your team is implemented.
   Please describe the structure or organization of your team.
   What is the team's official name and how many people are working within the team.
   What are the skills in your team and which ones would you consider most important.
   Are there any skill sets that you feel are missing on your team.

4. About your process
   How do you identify and choose your projects.
   Are there other important resources within your organization that you use for your work?
   What are the typical time frames for your projects.

5. About your communication channels

Which tools do you use to communicate within the team and with partners (e.g., cloud-based storage services, CRM, collaboration or networking tools).

Which tools do you use to manage projects, materials, and time frames.

Are there functions that you miss in those tools?

6. About your toughest challenges

Please describe the toughest challenge regarding the team's organizational management.

7. About your strongest recommendations

What are your three most important recommendations for newcomers in your field.

## INTERVIEWEES

| Name | Organization | Job Title |
|---|---|---|
| **Design Firms** | | |
| Duncan Clark | Kiln | Cofounder and Director |
| Audrée Lapierre | Function | Creative Director |
| Giorgia Lupi | Accurat | Cofounder and Design Director |
| Erik Jacobsen | Threestory | Practitioner of data visualization and the design of information |
| Duncan Swain | Beyond Words | Cofounder and Creative Director |
| Jason Lankow | Column Five | Cofounder |
| **Media Organizations** | | |
| Xaquin Gonzalez | *The Guardian* | Editor |
| Scott Klein | *ProPublica* | Deputy Managing Editor |
| Ulrike Köppen | BR Data | Team leader |
| Michael Kreil | *Tagesspiegel* | Data journalist and team leader |
| Julius Tröger | *Berliner Morgenpost* | Head of Interactive Team |
| Stefan Wehrmeyer | Correctiv | Data journalist |

*(Continued)*

| Name | Organization | Job Title |
|------|-------------|-----------|
| **Nonprofits and Nongovernmental Organizations** | | |
| Adam Coyne | Mathematica Policy Research | SVP of Communications, Chief of Staff |
| Pablo Picon Garrote | Inter-American Development Bank | Communications Senior Associate |
| Bridget Lowell | Urban Institute | Vice President, Strategic Communications and Outreach |
| Sarah Sattelmeyer | The Pew Charitable Trusts | Officer |
| Marie Whittaker | Office of the Deputy Mayor for Planning and Economic Development (DMPED) | Program analyst |

# CHAPTER 9

# Communicating Data to an Audience

**Steven Drucker**
*Microsoft Research*

**Samuel Huron**
*Institut Mines-Télécom*

**Robert Kosara**
*Tableau Software*

**Jonathan Schwabish**
*Urban Institute*

**Nicholas Diakopoulos**
*Northwestern University*

## CONTENTS

## INTRODUCTION

Communicating data in an effective and efficient story requires the content author to recognize the needs, goals, and knowledge of the intended audience. Do we, the authors, need to explain how a particular chart works? It depends on the audience. Does the data need to be traced back to its source? It depends on the audience. Can we skip obvious patterns and correlations and dive right into the deeper points? It depends on the audience. Do we need to explain what the findings in the data mean in terms of what the data represents? It depends on the audience. There are many more questions for which this is true.

It appears reasonable, then, to learn who that audience is and what they might know. In addition, designers might also want to know what their audience's expectations and needs are: what does the audience want to get out of the story? A single story cannot possibly address all possible different audiences and their needs. And any well-designed story will be tailored not just to its data and the intended message, but to its audience.

That is the theory, at least. In practice, this is quite difficult to achieve. A large audience will consist of people with different backgrounds and knowledge levels that are impossible to target at the same time. In breaking news and media production, short deadlines often make it impractical to create multiple versions of a news graphic that will work on different devices. In other fields, personnel or financial considerations may constrain the ability to create a product that targets the correct audience. Knowledge about the audience is also often quite limited. It tends to be general and broad, and usually not specific enough to target visualizations to individuals.

Despite these limitations, considering the audience, even broadly, can help guide the design of visual stories. The results are more appealing, effective, and meaningful to the intended recipients.

In some cases, content authors know precisely to whom a presentation or report will be given: what they know about the data, the background, etc. In other cases, the story author may at least have an expectation or an imagination of who they think the audience may be (Litt, 2012), such as the general level of data and visualization literacy of a particular publication's target audience. Such audience expectations inform the editorial decisions that need to be made about visualizations and news graphics. For instance, if the intended audience for a story is teenagers with limited statistical or visualization literacy, then additional explanation or context might be needed for less-common chart types so that they can be accurately interpreted.

Different audiences not only have different levels of familiarity and literacy with visualization, but they also have different goals, expectations, attention spans, and cognitive abilities. A general reader of a newspaper has very different needs and goals than an academic researcher reading a scholarly journal or a policymaker reading a briefing memo. A television audience will expect a different kind of presentation than a newsprint reader or a news website user, who in turn will have different expectations than those in a meeting with colleagues.

A television audience's attention span might be on the order of a few seconds. For Web audiences, some data show that 55% of users spend less than 15 seconds on an item of news content (Haile, 2014). By contrast, when communicating data and stories to people who need the information to make decisions or change policies—for example, executive managers or policymakers—their attention may be substantially longer because the information being communicated is essential to their performance. Other audiences may want to accomplish different things by reading a data story, such as being entertained, educated, or satisfying their general curiosity.

To effectively communicate ideas and concepts, content authors need to think carefully about how their work best fits the needs of the audience. In this chapter, we explore design considerations relating to audience knowledge and goal contexts, and consider the difference between the theory of what we might know and the reality of what we can know. We discuss some approaches that allow us to tailor a piece to the audience with little or no knowledge about them. Finally, we describe a few specific design contexts and their particular requirements, such as news graphics and broadcast television.

## WHAT DOES THE AUDIENCE KNOW?

Identifying the background knowledge of an audience for any given project is a challenging endeavor. In this section we identify some of the audience knowledge characteristics that are important when communicating information driven by data and told through text, images, and data visualizations.

Through discussion of our experiences coming from different backgrounds and past work experiences, we have identified three general areas where content producers could know more about their audience. Such knowledge would enable more effective communication of data-driven content.

1. How literate is the audience in terms of data and visualization? And what can be done to increase it?

2. How knowledgeable is the audience in terms of jargon, domain expertise, and other background knowledge?

3. What are audience expectations about the design of visualization, such as style, tone, or the use of iconography?

## Data and Visualization Literacy: The Annotation Layer

Definitions of visualization literacy describe the ability to interpret visual patterns of the underlying data, and to confidently use a visualization to answer questions about the data (Boy et al., 2014). As the value and availability of data continues to grow, so does the demand for understanding data and graphs—of increasing visualization literacy. Data science and data visualization courses and training in the private sector, in postsecondary schools, and in online training programs have increased substantially over the past few years (Womack, 2014). The onus is not just on the audience. Data storytellers have an ethical obligation to ensure that their visualizations do not skew or mislead interpretations unnecessarily (see Chapter 10, entitled "Ethics in Data-Driven Visual Storytelling"), and this intersects with the literacy level of the expected audience. Increasing an audience's understanding of different data and graph types may come from a variety of sources, some within the content-producers' control, others not.

In news graphics, the solution is what is usually called the *annotation layer.* Annotations add explanations and descriptions to introduce the graph's context, which is important for almost any audience (Hullman et al., 2013). They can also explain how to read the graph, which helps readers unfamiliar with the graph—whether a simple line chart or an advanced technique like a treemap or scatterplot. When done right, the annotation layer will not get in the way for experienced users.

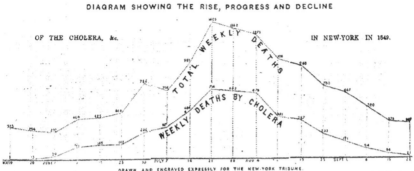

FIGURE 9.1   A line chart printed in the *New York Daily Tribune*, September 29, 1849 would not have been familiar to many of the newspaper's readers. As a result, extensive instructions are given for how to decode and understand the chart.

Annotations and explanations have a long history. Figure 9.1 shows a line chart in *The New York Daily Tribune* in 1849 in which the addition of an explanatory caption and annotations help guide the interpretation of how to read the then-still unfamiliar line chart.

Amanda Cox, the editor of the "Upshot" at *The New York Times*, is famously quoted as saying, "The annotation layer is the most important thing we do…otherwise, it's a case of here it is, you go figure it out" (Cox, 2011).

Annotation goes beyond just labeling points or lines on a chart: the bubble plot shown in Figure 9.2 from the *Los Angeles Times*, for example, expertly combines annotation that explains how to read the chart with what the content means. The chart shows the relationship between the change in violent crime rate (horizontal axis) and the property crime rate (vertical axis) in about 30 cities in California. For readers familiar with this chart type, it is immediately clear how to glean conclusions from the data. The average *Los Angeles Times* reader may not be familiar with this chart type, so upon first viewing the graphic, there is a big red box in the top-right with big red text that says "Worse," and a big blue box in

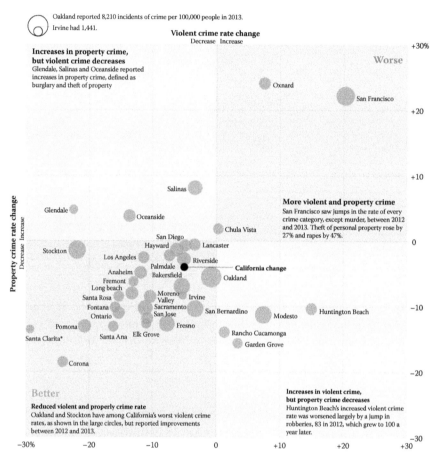

FIGURE 9.2   A bubble plot from the *Los Angeles Times* from 2013 expertly annotates both how to read the chart and the story the authors deliver.

the bottom-left with big blue text that says "Better." Then, each quadrant has a small headline in boldface type with a sentence below to deliver the content. Thus, even for the reader who has never seen a bubble plot before, the annotation instructs them on how to read the chart, and then delivers the content through the additional explanatory text.

## Background Knowledge and Expertise

A content producer may face an audience with different levels of expertise or knowledge on any given topic. Readers of a large daily newspaper like *The New York Times* or *The Washington Post* may reflect a variety of different levels of expertise, especially across the different topics covered by the publication. By contrast, readers of the *Financial Times* may have more domain-specific expertise in certain areas, especially as the *Financial Times* tends to focus their reporting on the financial sector. The audience for a small local publication is likely to have knowledge of the local area, whereas the same could not be assumed for the audience of a large national or international publication. Data storytelling needs to take into account the baseline of information and knowledge that the intended audience is expected or assumed to have.

Different levels of expertise and knowledge should encourage content producers to think carefully about how their products will best meet the needs of a diverse audience. Niche audiences may be more comfortable with the jargon of their group, but designers may consider whether removing jargon would serve a larger audience just as easily while making the content more broadly accessible. Universal design is an aspiration that can allow data stories to appeal to and be used by a broad array of people (Shneiderman et al., 2016), but design is also about tradeoffs that may make a data story less appealing to some, while simultaneously much more appealing and useful to others. In such cases, different versions of the story might be produced for different audiences, each targeting a unique range of knowledge, expertise, and other factors discussed in this chapter.

An individual in the audience brings their own viewpoints, backgrounds, and experience to each and every data-driven story they consume. Whether driven by their cultural background, social position, education, or other demographic characteristics, readers carry with them their own unique set of knowledge and biases. While it is impossible for content producers to be aware of each individual's particular experience— to know in advance what they already know—it may be possible to group individuals (i.e., to segment the audience) in ways that are useful to guiding design.

Some examples may be illustrative here. The first comes from the Congressional Budget Office (CBO), which is the budget arm of the United States Congress. The CBO reports and numbers are regularly used and referenced by Congress, as is often dictated by law. Their audience of the members of Congress and their staffs is well-defined and well-known. That audience wants (and needs) headline numbers, facts, and statistics that they can use to communicate with their colleagues and their constituents. In June 2012, the CBO published its "Long-Term Budget Outlook," a 109-page report about the budget outlook for the federal government (CBO, 2012). Paired with that report was a one-page infographic that highlighted the top-level items, facts, and patterns (see Figure 9.3). In a

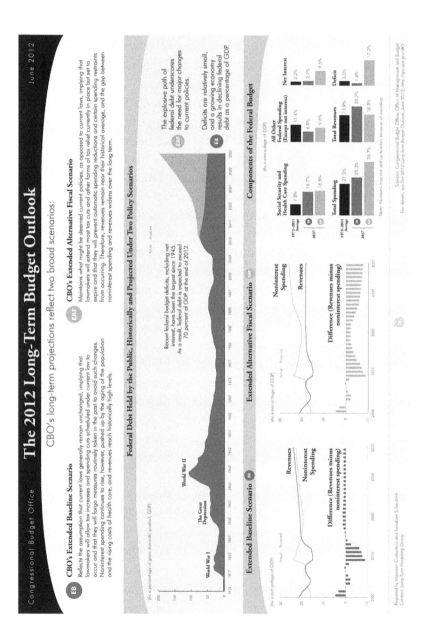

FIGURE 9.3    The 2012 Long-Term Budget Outlook infographic from the Congressional Budget Office.

congressional hearing about the long-term budget outlook, Ranking Member Congressman Chris van Hollen (D-MD) held up that infographic as an exhibit to support his point (CSPAN 2012). From the perspective of the CBO, this is arguably recognizing the needs of their audience and how a member of Congress could effectively use their analysis.

Another example comes from the Urban Institute, a nonprofit research organization in Washington, D.C. The Urban Institute is not only interested in communicating with policymakers, but is also interested in communicating their work to a broader audience of local and state policymakers, other researchers, practitioners, the press, and the public more broadly. In some ways, the Urban Institute is similar to the news organizations mentioned above, but communicating to an audience that is particularly interested in the kind of in-depth social and public policy research they conduct.

In August 2013, with the release of a report on the potential policy responses to long-term unemployment in the United States, the Urban Institute published a long-form narrative story on its website that combined the highlights from the report with interactive data visualizations, photographs, and sound clips from interviews with people who were unemployed (Acs, 2013). Recognizing that the traditional reports may not always have impact or reach a broad(er) audience, this type of data-driven story—which is now common for news organizations—is still fairly uncommon in the social science and nonprofit research space.

## Design Expectations

Basic data visualization best practices and strategies are needed across multiple domains to help content creators find better ways to communicate their work to their audiences. Many people who have experience working with data and conducting in-depth analysis may not have the knowledge or experience in how to *communicate* their work. The field of data visualization—a broad field that can be considered to encompass data journalism, data science, data visualization, scientific research, as well as general practitioners—continues to evolve in the face of new methods, tools, and data. A growing number of books, blogs, and articles have been published just over the past few years dedicated to helping people improve their visualizations.

A consistent design approach (one might call it "visual branding") can help an audience relate to a producer's work and entice them to return.

Consistency across color, font, layout, and type of interactivity, for example, can help the audience better identify what the producer is going to deliver, when, and for what level of expertise. Conventions used in presentations can also reinforce good data hygiene and inform the audience as to the quality of a visualization such as when source and designer credits or methodologies are consistently incorporated into the story delivery (Hullman and Diakopoulos, 2011; Schwabish, 2016a).

Graphics from outlets such as *FiveThirtyEight*, *Vox*, *The New York Times*, *The Economist*, and others each have a specific design aesthetic. The inclusion of their brands' colors and fonts, a specific text or annotation style, and method of publication (e.g., daily, blogs, etc.) all help audiences recognize and relate to the producers' content. Practically, a design aesthetic is enculturated by having style guides that designers working for a given publication use to help choose colors, fonts, and layout. Nonprofit organizations such as the Urban Institute* (see Figure 9.4) as well as news outlets such as the *Dallas Morning News*† have even published their style guides publicly. These guides help enforce a consistency of visual output (see Schwabish 2016b for a collection of other style guides).

Another dimension that informs design expectations is cultural differences and viewing codes. For instance, the layout of design elements such as the title and credits in Western cultures may not meet expectations in cultures that read from right to left. Such differences may be particularly relevant for complex designs that can evoke a layer of meaning via connotation. For instance, colors can suggest very different meanings in the East versus West: in Asia, white is often the color for mourning, whereas it is black in most of the West (Ware, 2008). This complicates the design process for data stories that are oriented towards a global audience such as may be the case for international publications.

## WHAT DOES THE AUDIENCE WANT?

Like any other designed experience, data stories are more impactful and useful when they are crafted with the motivations, goals, and tasks of users in mind—in this case the audience for the story. The following subsections explore several typical wants and needs from media consumption that may impact design, as well how interactivity can be leveraged to increase the adaptability of data stories to different user needs and contexts.

---

* http://urbaninstitute.github.io/graphics-styleguide/.
† https://knightcenter.utexas.edu/mooc/file/tdmn_graphics.pdf.

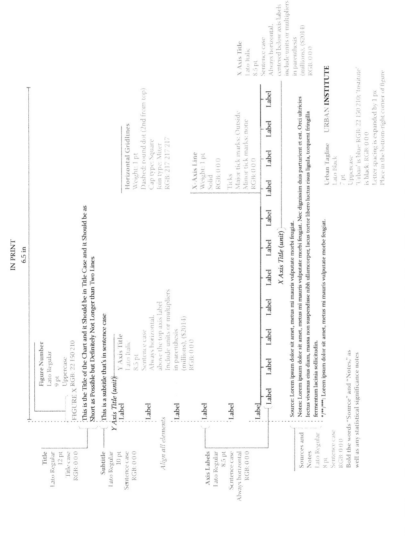

FIGURE 9.4    The "Chart Parts" section of the *Urban Institute Data Visualization Style Guide.*

## Media Wants and Needs

The uses and gratifications framework describes a set of dimensions that can influence why people consume media (Ruggiero, 2000). The idea is that people receive certain gratifications from the media they consume that help satisfy various underlying social or psychological needs. For instance, people may seek out data-driven stories for a variety of reasons such as to inform themselves on an issue of personal (e.g., health) or civic importance (e.g., elections), or to casually lean back and be entertained or pass time in a non-goal-oriented fashion. The uses and gratifications framework suggests that there are four primary underlying motivations for media consumption:

1. Staying informed about current information, satisfying curiosity and self-education;

2. Gaining insight into oneself or seeing models for others' behavior;

3. Socially interacting and connecting with others, or as a basis for social conversation or developing empathy; and

4. As relaxing entertainment or diversion.

A data-driven story can be designed differently depending on whether the audience is expected to approach the experience from a more social, entertainment, or strictly informational motivation. For example, an entertaining treatment might take more license with visual novelty and not focus as closely on efficiency in decoding visual information. A social treatment might emphasize user comments and responses to a data story, whereas an information-oriented design may focus on well-researched and validated background context provided in an annotation layer. The goal as a designer is to explore the design space in order to try to match the data story to the expected media need of the audience.

An individual's needs from a piece of media like a data story can also vary with the intended interface. For instance, if a story is likely to be consumed on a mobile device, the user could be in a loud or distracting environment that might undermine the use of audio in the narrative (written annotations as backup is a nice design strategy). The specific location of a user on a GPS-enabled device can be interesting context that can be used to inform the way a story is presented (Kim et al., 2016). In contexts that are more casual than work-oriented, users may have a higher degree of interruptibility, which could mean that a data story is consumed in short bursts of attention. Presenting information in "chunked" patterns might be more effective than as a single, long-form narrative in such cases. Finally, it is worth considering that boredom may be a driving factor for interaction in contexts where the user is consuming a data story while waiting (e.g., for a bus or train), or splitting attention between devices and screens (e.g., as many people do when watching television).

## Tailoring to the Audience without Knowing It

Even if we do not know anything about the audience, we can make it appear as if we did. The audience knows itself: they can pick their city in a story about home prices, their age and gender in a story on health, and their education and income levels in a story about taxes or inequality.

This self-identification in a visualization may be why maps are such a popular tool in data visualization. While maps can often present data inaccurately, or even mislead a reader, broad familiarity with maps makes them a convenient and familiar visualization choice (Wiseman, 2015). Alternatives—such as cartograms, which distort geographic areas to correspond to the data values—sometime have the advantage of portraying the data in more accurate ways, but the distortion makes the visualization less familiar to the reader.

A variety of other projects and research have tested or demonstrated the importance of allowing the audience to "find themselves" or make content more familiar in a visualization. For example, research by Chevalier et al. (2013) attempts to build a framework to help designers think about the best ways to communicate numbers that are very large. More specifically, Kim et al. (2016) construct an online interactive application that enables users to re-express spatial measurements in localized terms—in other words, users are able to build a map that makes a distance relevant to their area or experience.

Even outside of maps, readers locating themselves in the data is likely to make them more engaged. A piece breaking down the unemployment rate by race, gender, age group, and education level (Carter et al., 2009) allows the reader to compare themselves to the average and to other specific groups. While not all possible criteria are covered, this piece gets much closer to an individual than just the unemployment rate that averages across everybody. Similarly, a line chart showing the Case-Shiller Home Price Index for 20 cities lets the reader pick their city and see how housing prices have changed over the years (Carter and Quealey, 2014).

### Directly Engaging the Audience

A somewhat similar approach asks the audience questions to involve them in what is shown. This can help engagement by allowing them to place themselves in a context (and at the same time also collect data to be shown to everybody), and it has also recently been found to help people remember what they have seen.

In 2011, *The New York Times* published "The Death of a Terrorist: A Turning Point?" (Huang and Pilhofer 2011) where readers were invited to position their opinions on the death of Osama bin Laden in a two-dimensional matrix: whether their response was positive or negative overall, and whether they thought that this event was going to be significant in the war on terror or not. The resulting heatmap of responses showed the landscape of people's responses.

Analogous pieces have since appeared in different media, such as *The New York Times'* "You Draw It: How Family Income Predicts Children's College Chances," which allowed users to draw their idea of the correlation between parents' income and their children's ability to attend college (Aisch et al., 2015). Similar data stories have been published to let readers predict price changes for common goods to match an inflation goal (Canipe et al., 2016), draw the border between West Germany and the German Democratic Republic from memory (Keseling et al., 2015), or guess the rate of teenage pregnancies in the United Kingdom (ONS, 2016) (Figure 9.5).

Interactive pieces like this are still relatively rare, but they are slowly gaining momentum. A recent study tested the effect of asking readers to make estimates before showing the actual data on memory (Kim et al., 2017) (Figure 9.6). It showed that asking participants to reflect on their prior knowledge by guessing numbers or explaining them helped readers with both comprehension and memory of the data.

Engaging an audience is possible even without much knowledge about its members' backgrounds, level of domain knowledge, etc. Personalization as discussed in the previous section and engagement through questions have both been tried and found to be quite effective.

## SPECIFIC DESIGN CONTEXTS

In addition to the audience itself, the medium used to reach that audience is an important factor that shapes what is possible in practice and what people expect. A detailed news graphic designed for print will not work on television, and a television piece will look out of place on the Web.

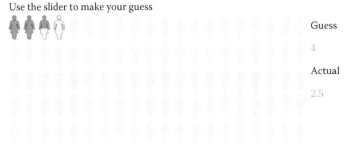

**For every 100 girls aged under 18 in 2014, how many become pregnant?**

This is all conceptions for girls under 18, per 100 girls aged 15 to 17

Use the slider to make your guess

Guess

4

Actual

2.5

You're not too far away but you've gone a bit too high!

FIGURE 9.5 In this interactive piece by the UK's Office of National Statistics, readers are first asked to make an estimate using the slider at the bottom, which fills in as many people as they are guessing (girls under 18 becoming pregnant, in this case). When they submit their guess, the difference between it and the real value is illustrated by a small animation and the numbers are shown on the right (ONS, 2016).

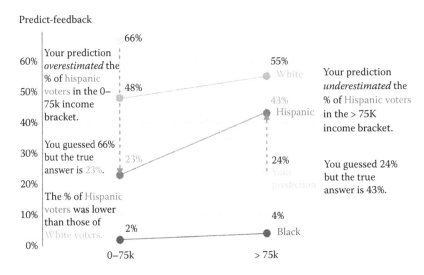

FIGURE 9.6 User interface used in a study to test the effectiveness of engaging readers by asking them to make guesses about the data first and then showing them how close (or far off) they were (Kim et al., 2017).

## The Reality of the Newsroom

Over the last several years, the news media have created many of the most exciting examples of data-driven storytelling. News media, like other organizations, have struggled with different device types, browser idiosyncrasies, screen sizes, as well as different technologies and programming languages (such as the switch from Flash to HTML5).

Designing for devices means not only taking their idiosyncrasies into account, but taking advantage of them. Trying to design across many devices typically ends up with a sort of lowest common denominator that works but is not as effective as it could be on any specific device. See Chapter 6: "Watches to Augmented Reality: Devices and Gadgets for Data-Driven Storytelling" for more details.

Recently, many newsrooms have scaled back those efforts. *The New York Times* Graphics Editor Gregor Aisch described the problem as a cost-benefit tradeoff (Aisch, 2016). Many interactive news pieces see very little interaction—only 10%–15% of visitors interact with them. Even a simple secondary interaction, like left-right swiping to reveal more information when the main flow of the story is swiping down, is rarely used. This is despite efforts like adding prompts.

Designers have devised a number of work-arounds to make large graphical pieces work on small screens. For maps, a convenient one is to zoom the map into the location of the user, since that is likely what they are interested in. However, that often means that there is no way to zoom out to see it in its entirety. Aisch therefore proposes a variation of Shneiderman's famous information-seeking mantra: "*Overview first, zoom and filter, then details on demand*" (Shneiderman, 1996) and for the reality of news graphics on mobile devices: "*Details first, no zoom and filter, overview on desktop only.*"

Newsrooms' resources are limited, and they often produce news stories on tight deadlines (see also Chapter 8: "Organizing the Work of Data-Driven Visual Storytelling"). In light of this, it becomes necessary to consider not just the potential information value, but also the effort to make something interactive when so few readers ever get a benefit from it. Similar to Aisch, Tse (2016) argues that interactive data stories (at least in the newsroom) may not be the best approach. He makes a three-point argument:

1. If you make the reader click or do anything other than scroll, something spectacular has to happen.

2. If you make a tooltip or rollover, assume no one will ever see it. If content is important for readers to see, do not hide it.

3. When deciding whether to make something interactive, remember that getting it to work on all platforms is expensive.

Tse's argument rests on changes in technology, and the changing business structure of the media industry (e.g., advertising efforts, subscription packages, etc.).

It is not clear what the solution is to these issues. One approach is certainly better tools that make it easier to create pieces in the first place and allow the author to target different platforms, either automatically or at least with some support. Another is to show potential interactions with little animations. This has been tried in projects such as Gapminder, but we are not aware of any studies that show if such animations actually increase interaction. Perhaps the most important issue is still training the consumer of the information: if readers are not used to charts responding and allowing them to interact, they will not try hovering over elements, etc. By creating more interactive pieces, media can train their readers to expect everything to be interactive, and thus increase interaction over time (whether this works or not will have to be studied longitudinally).

While news media often have a general idea of their audience, it tends to be vague and hard to use for practical decision-making. There are still considerable hurdles to using what is known to target the audience more specifically and precisely. There are many open questions here and ample need (and opportunity) for more research that could help improve the way news pieces are built and how they work.

## Visualization for Television

Television is an unusual use case for visualization, even though its visual nature seems to lend itself well to it. This case study is based on the program *Le Grand Webzé*, which aired on February 24, 2012 on the France 5 channel (Huron et al., 2013b).

A television audience is highly diverse: almost any age, any level of education, any social position, economic role, and cultural interest is represented. Ratings agencies have been tracking audiences for decades, with age groups starting as young as 4 years old, with segments by age, gender, economic status, and if they are living in urban or rural areas. Television audiences are also fickle, often switching programs and channels within less than a second.

The research and development services of the French National Television Network contacted Huron and his colleagues to create a system that would engage a television audience with online activity during a television show. The program is focused on Web culture and presents a review of the latest Web buzz as well as interviews of famous bloggers or Web artists. The goal was to allow viewers to interact with each other and to see a visual representation of that interaction on the program.

The design was made for the medium and the way it is consumed. To show vote tallies, bubbles would drop and form into bar charts (see Figure 9.7). The bubbles included small images of the people who had voted, making for a selfie moment and improving the visual interest of the graphic. Animation was used throughout to blend in with the always-moving nature of television.

Television presents unique design challenges for visualization: the amount of detail is very limited, there is no interaction, motion and animation are crucial, television graphics are often three-dimensional, etc. Despite its prevalence in popular culture, television as a medium is still almost completely overlooked in visualization research.

FIGURE 9.7 Photograph taken of the studio monitoring screen during the *Le Grand Webzé* television program. The visualization shows the votes of the show audience about the host's popularity. Each circle represents a tweet sent during the last few seconds, the bar chart represents the total number of tweets for each hashtag. The visualization is updated continuously.

## CONCLUSIONS

When presenting data and telling stories, it seems obvious to tailor the presentation to the audience. In practice, unfortunately, there are many reasons why this is not easy or even possible. It could be due to financial or personnel constraints, deadlines, uncertainty about the targeted audience, or even ignorance about audience capacities. Despite such uncertainty, one way to think about the audience could be to consider their data and visual literacy level, their attention span, background knowledge, and domain expertise.

The inability to know enough about the audience in many cases has led to some interesting approaches that have proven to be helpful beyond just tailoring a visualization to the reader or viewer. Personalization is an effective means not only to tailor a news piece to a particular reader in more detail than could be done otherwise, but also a way to help engage that reader. Asking for the person's background knowledge by having them estimate a value first is necessary when the visualization designer does not know the reader, but this has proven helpful with memory and comprehension of the data being shown as well.

In an ideal world, the designer of a data story would know all the relevant information about his or her audience and be able to put that into practice. Some things are easier to measure or assess, like a person's visualization literacy or familiarity with a particular chart type. Others are harder or close to impossible, like a person's background knowledge about a topic or question they might have. Even without being able to answer these questions, a well-designed story brings in many different viewers by providing pieces of context that are likely relevant to most users without cluttering up the display. Going further, it is conceivable that a story would measure a viewer's visualization literacy in an unobtrusive (or perhaps indirect) way and adapt the amount of information being shown, as well as the types of visualization, etc.

Smart design takes available knowledge about the audience into account, but does not depend on it being available. Many design decisions are also driven more by general ideas about effective presentation than specifics about a particular audience. A better understanding of the audience will likely lead to better data stories, but the current state of the art in news graphics, television, and other areas shows that good designers are able to create compelling stories even when faced with a limited knowledge of the actual audience.

## REFERENCES

Acs, Gregory. 2013. Responding to Long-Term Unemployment, Urban Institute Report.

Aisch, Gregor, 2016. Data Visualization and the News, presentation at the 2016 Information Plus Conference. https://vimeo.com/182590214.

Aisch, Gregor, Amanda Cox, Kevin Quealy. 2015. You Draw It: How Family Income Predicts Children's College Chances, *The New York Times*. https://nyti.ms/2jX8zue.

Boy, Jeremy, Ronald A. Rensink, Enrico Bertini, Jean-Daniel Fekete. 2014. A Principled Way of Assessing Visualization Literacy. *IEEE Transactions on Visualization and Computer Graphics*, 20 (12), 1963–1972.

Canipe, Chris, Katie Marriner, Stuart A. Thompson, Andrew Van Dam. 2016. Beat the Federal Reserve: See How Changes in Prices of Goods and Services Influence Inflation, *The Wall Street Journal*. http://www.wsj.com/graphics/beat-the-fed/.

Carter, Shan, Amanda Cox, Kevin Quealy. 2009. The Jobless Rate for People Like You, *The New York Times*. http://www.nytimes.com/interactive/2009/11/06/business/economy/unemployment-lines.html.

Carter, Shan, Kevin Quealy. 2014. Home Prices in 20 Cities, *The New York Times*. https://nyti.ms/2kMCid7.

Chevalier, Fanny, Romain Vuillemot, Guia Gali. 2013. Using Concrete Scales: A Practical Framework for Effective Visual Depiction of Complex Measures. *IEEE Transactions on Visualization and Computer Graphics*, 19 (12), 2426–2435.

Congressional Budget Office. *The 2012 Long-Term Budget Outlook*. June 5, 2012.

Cox, Amanda. 2011. Shaping Data for News, presentation at the 2011 Eyeo Festival. https://vimeo.com/29391942 (quote at 19:29 seconds).

CSPAN. 2012. Long-Term Budget Outlook, video. https://www.c-span.org/video/?306443-1/longterm-budget-outlook, June 6, 2012.

Haile, Tony. 2014. What You Think You Know about the Web Is Wrong. Time. http://time.com/12933/what-you-think-you-know-about-the-web-is-wrong/.

Huang, Jon and Aron Pilhofer. 2011. The Death of a Terrorist: A Turning Point? *The New York Times*. http://www.nytimes.com/interactive/2011/05/03/us/20110503-osama-response.html.

Hullman, Jessica, Nicholas Diakopoulos, Eytan Adar. 2013. Contextifier: Automatic Generation of Annotated Stock Visualizations. *Proceedings of the SIGCHI Conference on Human Factors in Computing Systems (CHI)*, ACM, New York, pp. 2707–2716.

Hullman, Jessica, Nicholas Diakopoulos. 2011. Visualization Rhetoric: Framing Effects in Narrative Visualization. *IEEE Transactions on Visualization and Computer Graphics*, 17 (12), 2231–2240.

Huron, Samuel, Romain Vuillemot, and Jean-Daniel Fekete. 2013a. Visual Sedimentation. *IEEE Transactions on Visualization and Computer Graphics*, 19 (12), 2446–2455.

Huron, Samuel, Romain Vuillemot, Jean-Daniel Fekete. 2013b. Bubble-TV: Live Visual Feedback for Social TV Broadcast. *ACM CHI 2013 Workshop: Exploring and Enhancing the User Experience for Television*, April 2013, Paris, France.

Keseling, Uta, Max Boenke, Reto Klar, Julius Tröger, Christopher Möller, David Wendler, Moritz Klack. 2015. Wissen Sie noch, wo Deutschland geteilt war? (Do you still remember where Germany was separated?) Berliner Morgenpost. http://einheitsreise.morgenpost.de.

Kim, Yea-Seul, Jessica Hullman, and Maneesh Agrawala. 2016. Generating Personalized Spatial Analogies for Distances and Areas. *Proceedings of the 2016 CHI Conference on Human Factors in Computing Systems*, pp 38–48, San Jose, California.

Kim, Yea-Seul, Katharina Reinecke, Jessica Hullman. 2017. Explaining the Gap: Visualizing One's Predictions Improves Recall and Comprehension of Data, *Proceedings of the 2017 CHI Conference on Human Factors in Computing Systems*, pp. 1375–1386. Denver, Colorado.

Litt, Eden. 2012. Knock, Knock. Who's There? The Imagined Audience. *Journal of Broadcasting & Electronic Media*, 56(3), 330–345.

Office for National Statistics (ONS). 2016. Teenage Pregnancies— Perception versus Reality. http://visual.ons.gov.uk/teenage-pregnancies-perception-versus-reality/.

Ruggiero, Thomas E. 2000. Uses and Gratifications Theory in the 21st Century. *Mass Communication & Society*, 3 (1), 3–37.

Schwabish, Jonathan. 2016a. *Better Presentations: A Guide for Researchers, Scholars, and Wonks*. Columbia University Press, New York.

Schwabish, Jonathan. 2016b. Style Guides, blog post. https://policyviz.com/2016/11/30/style-guides/, November 30, 2016.

Shneiderman, Ben, Catherine Plaisant, Maxine S. Cohen, Steven Jacobs, Niklas Elmqvist, Nicholas Diakopoulos. 2016. *Designing the User Interface: Strategies for Effective Human-Computer Interaction* (6th Edition). Pearson. Essex, UK.

Shneiderman, Ben. 1996. The Eyes Have It: A Task by Data Type Taxonomy for Information Visualizations. *Proceedings of the IEEE Symposium on Visual Languages*, IEEE, Washington, DC, pp. 336–343.

Tse, Archie. 2016. Why We Are Doing Fewer Interactives, presentation at the 2016 Malofiej Infographics World Summit. https://github.com/archietse/malofiej-2016/blob/master/tse-malofiej-2016-slides.pdf.

Ware, Colin. 2008. *Visual Thinking for Design*. Morgan Kaufman Publishers, Burlington, Massachusetts

Wiseman, Andrew. 2015. When Maps Lie, CityLab. http://www.citylab.com/design/2015/06/when-maps-lie/396761/, June 25, 2015.

Womack, Ryan. Data Visualization and Information Literacy. *IASSIST Quarterly* 2014.

# CHAPTER 10

# Ethics in Data-Driven Visual Storytelling

## Nicholas Diakopoulos

*Northwestern University*

## CONTENTS

The Accountable Journalism Project at the University of Missouri touts having compiled a database of more than 400 codes of media ethics from around the world.* Yet a topical search of the codes tagged with "data

---

* https://accountablejournalism.org.

journalism" yields only 17 results—that is only about 4%. Well-regarded ethics codes like those from National Public Radio (NPR)[*] and the Associated Press (AP)[†] do at least mention data and graphics, and a revised 2017 version of the *AP Stylebook* includes a whole chapter on best practices in data journalism, but, for example, a widely used ethics guide from the Society for Professional Journalists (SPJ) does not mention data at all and includes only an oblique reference not to "distort...visual information."[‡] Despite the use of data-driven graphics in the news context for at least a couple hundred years, a robust articulation of ethical mandates for responsible usage has largely failed to emerge and become widely adopted.

This chapter is a step in the direction of articulating a framework for thinking about ethical decisions and pitfalls that relate directly to creating data-driven visual stories. In particular, I approach this within the pragmatist-realist discourse on data visualization (Dick 2016), meaning that I focus on the utilitarian usage of data visualization in the journalistic news media in comparison to other usages for persuasion or for artistic storytelling, which have different ethical commitments.

Ethics is concerned with articulating apt behavior and conduct: how ought one act according to standards of character? Journalism ethics scholar Stephen Ward defines it as "the study and practice of what constitutes the best regulation of human conduct, individually and socially" (Ward 2015). In the context of this chapter, ethics is about making sound editorial decisions along each step of the data-driven visual storytelling process: from the collection and acquisition of data to the analysis and presentation of information.

Principles can be useful guideposts for generally applicable beliefs about ethical behavior. It is important to articulate principles and delineate how they apply because having clearly described codes of ethics can have beneficial impacts on professionals' behavior (Powell and Jempson 2014). Here I do not articulate new principles but draw on those proffered by McBride and Rosenstiel (2013) and examine how they apply to the unique and specific demands of data-driven story production. These principles include:

1. "Seek truth and report it as fully as possible."

2. "Be transparent."

3. "Engage community as an end, rather than as a means."

---

[*] http://ethics.npr.org/.

[†] http://www.ap.org/company/News-Values.

[‡] http://www.spj.org/ethicscode.asp.

Principle #1 is particularly relevant to data stories because of the many possibilities to mislead, misinform, obfuscate, or even deceive using visualization. As will be discussed in the next sections, opportunities for (mis-) guiding end-users' interpretation of data can arise during every phase of production from acquisition and how the data is quantified, to how it is normalized, aggregated, filtered, visually encoded, annotated, and made interactive.

Principle #2 is relevant to the ways in which insights from data are found and interpreted as stories. Given that many interpretations may arise from a data set, it becomes important for the storyteller to be able to trace and disclose their rationale and process for arriving at a preferred interpretation. This is particularly important as it relates to data acquisition and transformation steps that may go unseen unless they are given explicit consideration for disclosure.

Finally, principle #3 hints at another dimension of importance when telling stories with data: the individual people who the data represents are to be treated with humanity and respect and not be seen only as a means for telling a story. In terms of supporting the goal of community engagement, this principle also connects to interactivity in data-driven storytelling. Designers can engage community through interactivity with the data rather than dictating data as if it is immutable.

Besides moral righteousness, acting ethically in data-driven visual storytelling can have positive outcomes for end-users. Alberto Cairo has addressed ethics in this domain by espousing a utilitarian perspective primarily focused on the end-user benefit of improved understanding and knowledge through the reception of accurate and compelling information (Cairo 2014a, 2017). Acting ethically in this domain entails not only honesty and virtuous intent but also considering and minimizing the potential errors of interpretation that may ultimately mislead viewers (Cairo 2014b; Bradshaw 2015). The utilitarian perspective is worth embracing, and expanding to consider the social benefit of increased trust of individuals or organizations acting ethically by adhering to the principles listed above.

In the next sections I examine ethical considerations that may arise across the data-driven story production pipeline: in data acquisition, in data transformation and interpretation, and in the design and conveyance of those insights for consumption. An understanding of ethical visualization design is further informed by considering how designers can use rhetorical techniques to strategically prioritize certain interpretations

(Hullman and Diakopoulos 2011). The goal is to describe why each ethical consideration is important and relevant to the principles articulated above, providing illustrations from concrete cases where possible. I strive to provide a range of ethical considerations that designers may encounter while acknowledging that this single chapter cannot possibly cover all of the myriad ways in which ethics touches data-driven visual storytelling.

## DATA ACQUISITION

### Provenance

The origin of the data used, including the possible political or advocacy motives of the data provider are essential to understand when striving to tell informative and truthful data narratives. Who produced the data and what was their intent? Is it complete, timely, and accurate? It is important to provide a baseline of transparency by citing or linking to the data sources used. But there is also the ethical question of whether to use a data set at all, such as if the progenitors of the data may have a stake in misinforming, distracting, or promoting a version of the truth that may be at odds with the public interest, or if the data was obtained illegally such as through hacking or an anonymous leak.

"The wave of bullshit data is rising, and now it's our turn to figure out how not to get swept away," writes Jacob Harris, a former *New York Times* software developer.* So-called "PR data" is released by private organizations with the intent of positioning their own organization favorably, or to gain free brand awareness by being listed as a data source on a story that gets a lot of attention. For instance, a story published by *The Washington Post* (and similar stories from many other outlets) used data from the pest company Orkin to present a ranking of the "top 20 rattiest cities" according to where Orkin made the most number of rodent treatments.† The numbers suggest the rat problem in Washington, D.C. is bad, worse than New York City even, but it is unclear whether given the bias in the data collection that this is unnecessarily stoking concern for a "rat problem" in the city when it may not really exist. For instance, 2013 data from the U.S. Census American Housing Survey ranked Washington, D.C. as 13th in terms of proportion of occupied housing with reported rat sightings.‡

---

* http://www.niemanlab.org/2014/12/a-wave-of-p-r-data/.
† https://www.washingtonpost.com/blogs/local/wp/2014/10/13/rats-d-c-calls-pest-company-about-rodents-more-often-than-new-york/.
‡ http://www.bloomberg.com/news/articles/2015-07-30/these-are-the-most-vermin-filled-cities-in-the-u-s-.

Even for ostensibly reputable sources like governmental or government-funded organizations, it can be important to understand why the data was collected to begin with, how politics may have affected its collection, and whether then it is still suitable to repurpose the data for a given storytelling context. For instance, data journalist Nicolas Kayser-Bril suggests politics may color how the Global Terrorism Database defines and tabulates acts of violence in Eastern Ukraine as terrorism rather than as acts of war between Ukrainian and Russian militaries.[*]

## Quantification

The way in which the world is measured and turned into data is another ethically interesting consideration. This includes not only how the world is sampled and how individual data are measured, but also how measurements are defined to begin with, and where to draw the line between something potentially ambiguous that is counted as category "A" rather than category "B." This can be further complicated when different stakeholders have different definitions. As we put a ruler to the world we cannot help but lose information and context in the creation of data, and so it is important to consider whether what is measured in the data allows for a given story to be faithful to reality. Are we really measuring what we mean to measure?

Whether counting ethnicity, mass shooting events, education, or unemployment, the definition of "what counts" in a particular category matters (Stray 2016). An example of the power of definition comes from the Chicago police department, which from 2010 to 2013 managed to reduce the rate of 8 index crimes by a staggering 56%. But an investigative report from *Chicago Magazine* suggests that these numbers were too good to be true and that one of the ways the stats had been "rinsed and washed" was to misclassify and downgrade offenses so that they were not tallied as part of the eight index crimes.[†] To tell an honest story about the changes in crime, it is essential to know the exact definitions of what is tracked and how those definitions could be subjectively massaged (or simply misunderstood or misapplied) by the data collector, or be legitimately changed over time. In this case, understanding the social and political pressures on the crime quantification process in Chicago led to an extensive story of its own.

---

[*] http://blog.nkb.fr/data-free.
[†] http://www.chicagomag.com/Chicago-Magazine/May-2014/Chicago-crime-rates/.

## DATA TRANSFORMATION

### Normalization

XKCD comic creator Randall Munroe illustrates the core problem with non-population-normalized mapping in Figure 10.1. The basic issue is that, for many types of data, plotting them on a choropleth map will simply show hot spots in areas of high population. The map ends up showing high population density around urban areas rather than the variable of interest. The correction here is to divide by the population in the area (e.g., state) that is being mapped, thus transforming the variable mapped to a population-normalized value.

Of course, normalization should not be done blithely. The ethical decision here concerns how choices in normalization affect the insight or story that a reader will come away with, including the clarity with which they see that story. For instance, in charting variables that typically have a large exponential skew, such as social media retweets or likes on posts, a

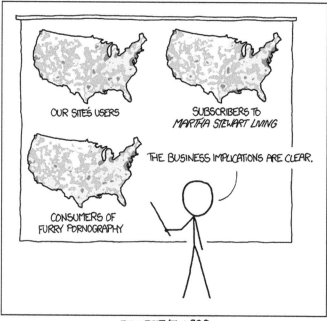

FIGURE 10.1 XKCD comic points out how non-normalized maps often just indicate population density.

visualization designer might choose to log-transform the variable to make the range of values more visible. In charting variables relating to the economy over time it is often necessary to adjust and normalize for inflation so that the underlying signal is faithfully shown.

One way to manage the ethical decision of normalization is to give the end-user more freedom to make this decision on their own terms. Interactivity can facilitate exploration of different types of normalization and their effect on the take-away of a story. An interesting example of this is Google's "Alternative Olympics medal table," which shows the rankings by country of the various Olympic medals in the 2016 games.* At the top of the interface are options to normalize the results by population, gross domestic product (GDP), and other factors to see how the various countries compare. The presentation draws attention to the fact that the rankings will differ when normalized in different ways, adding to the complexity and depth of the data presentation while putting the ethical decision of normalization in the hands of end-users.

## Aggregation

When telling a story with aggregated data—data that is derived through mathematical operations or combinations—it is ethically preferred to ensure that a consistent form of aggregation is applied to any data that are to be visually compared. This includes basic methods of aggregating a set of numbers such as taking the mean or median, as well as other operations like binning values to create a histogram. For instance, bin widths for aggregating values into counts for a histogram should ideally be made consistent so that visual comparisons are made according to commensurate aggregations. When aggregating different data measures through operations such as summation, it is important to ensure that the definitions are compatible so that the summation is meaningful and the combined value is easily interpretable.

Consider the chart from *The Economist* in Figure 10.2, which shows historic and forecast vehicle sales from different world geographies. The sharp upward trajectory of the line for China tells a clear story: sales in China are moving up quickly while they are declining or stagnant in North America and Europe. But an important caveat here is that the data for China (and India) has been aggregated differently than for the other geographies depicted—it includes truck, SUV, and minivan sales whereas

---

* https://landing.google.com/altmedaltable/results/.

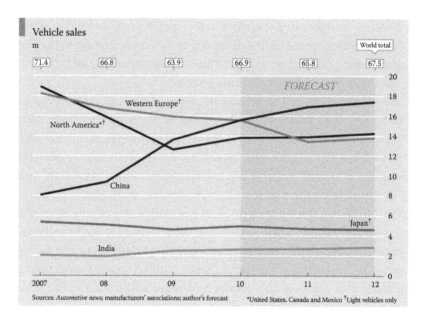

FIGURE 10.2   Historic and forecast vehicle sales around the world. The visual comparison here is misleading because the different data series are defined and aggregated in different ways.

the other geographies exclude those sales and only aggregate light vehicle sales. While the graph does have a small note indicating "light vehicles only" and is transparent about how it has aggregated the data, many readers may simply glance at the chart and see the visual impression of a quickly growing China in comparison to other geographies.

## Algorithmic Derivation

In addition to normalization and aggregation, more sophisticated algorithms are sometimes employed to derive new pieces of data. For instance, an analyst might use a classifier to predict a social media user's age or sex based on their connections or posts, or a sentiment analysis routine to determine whether an online review is positive or negative in tone, or a geocoder to translate a written address into a latitude/longitude pair. But any of these algorithmic processes can introduce errors into data and it is important to consider how that might impact the intended story. In 2009, the Los Angeles Police Department was notified of geocoding errors in its public crime mapping portal.[*] When the geocoder could not locate the

---

[*] http://www.latimes.com/local/la-me-geocoding-errors5-2009apr05-story.html.

address of a crime, it would default to an arbitrary location in Los Angeles which made it look like there was a crime hotspot downtown when in fact this was just the result of errors accumulating. When employing algorithmically derived data in visual stories, it is necessary to understand the nature of the errors introduced and try to mitigate the effect those errors might have on end-users' understanding. This can be done by communicating errors or uncertainties in the data via transparency practices such as methods sidebars or by open-sourcing the methods and data (Stark and Diakopoulos 2016).

## Filtering

Filtering what is shown is an essential operation that enables successful data-driven storytelling by focusing attention on relevant cases and removing distraction. Filtering data, or the visual process of cropping, requires ethical consideration with respect to the relevance (or irrelevance) of variables, individual cases, or ranges of variables being visualized. In some cases, outliers may be removed from the display (or removed before an aggregation step). Filtering also relates to the idea of cherry-picking—selecting only the data to display that supports an opinion the author has or which otherwise benefits the author in some way. Intentional cherry-picking is surely unethical, but even unintentionally filtering away data that forms essential context or comparison for nonfiltered data could end up being misleading. An ethical question worth engaging is whether the main takeaway from a visual is significantly altered as a result of any filtering that has been applied to the data or to the visual field.

## Anonymization

Personally identifying information such as names and addresses is sometimes present in data sets that may underpin a data-driven story. When names and addresses are coupled with other personal information such as political views, arrest or criminal records, or even just consumer behavior (e.g., buying a gun), this can lead to situations where personal privacy issues may come into tension with publishing in the public interest (Bradshaw 2015). While it might be considered an issue of public interest if an official figure like a mayor has an arrest record, it is unclear that having this information for an individual not in the public eye is newsworthy.

Anonymization can be a useful data transformation that protects individual privacy by filtering a data set of any personally identifiable traits that might allow an end-user to infer any identities from the data presentation.

In some cases, anonymization may trigger a need for a change in granularity of aggregation or visual presentation. For instance, consider a map published in 2012 by *The Journal News* in Westchester, NY showing dots at each address where there was a gun permit holder. After publication there was considerable pushback and concern that the map made it too easy to see personal information that could create risks for individuals. Because members of a community can easily infer the identity of each dot based on the address shown, complete anonymization would require a reduction in visual fidelity such as by visualizing the data aggregated by zip code.*

## CONVEYING AND CONNECTING INSIGHTS
### Visual Mapping and Representation

Edward Tufte's advice on graphic integrity still rings true: "A graphic does not distort if the visual representation of the data is consistent with the numerical representation" (Tufte 2001). There are a number of ways in which data can be mapped to distort and mislead the viewer, and which the ethical communicator must learn in order to avoid.

The way in which data is mapped spatially along a set of axes is an important decision in conveying an insight. Set the range of an axis too small and a trend or spike could get squashed, effectively hiding it from the viewer. Set the range too high and that same trend or spike could give an exaggerated and disingenuous impression. Sometimes it is appropriate to start an axis at zero (e.g., for bar charts because values are read according to the length of the bar), whereas in other cases (e.g., for dot plots or line charts), a nonzero axis starting point is acceptable and can even help clarify nuances to the data. In addition to truncating an axis, inverting an axis is another mechanism that can mislead (Pandey et al. 2015). By flipping an axis, a design might cause an end-user to read an increase as a decrease and vice versa.

Figure 10.3 shows an example of a truncated y-axis that is misleading. In moving from a value of 75% in 2008–2009 to 78% in 2009–2010, the height of the bar (represented as a stack of books) is doubled, visually exaggerating the actual increase. Another misleading aspect of this chart is that the height of the bar from 78% to 79% goes up quite a bit, but for the exact same increase from 81% to 82%, the bars are barely different in

---

* http://www.poynter.org/2012/where-the-journal-news-went-wrong-in-publishing-names-addresses-of-gun-owners/199218/.

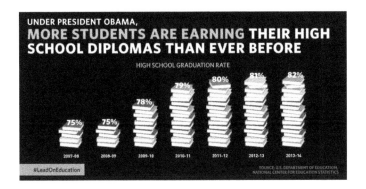

FIGURE 10.3   A misleading chart tweeted by the White House uses a truncated y-axis and inconsistent mapping between bars.

height. In both cases, the height of the bar should go up the same amount so that the data to visual mapping is always consistent.

Other choices in visual mapping and representation can also mislead viewers. A user study of deception techniques in data visualization showed that stretching a chart and changing its aspect ratio can influence how severely a trend is perceived (Pandey et al. 2015). A classic mechanism for deception is to map a value to a 2- or 3-dimensional shape, which distorts the representation because as the value increases linearly the area or volume of a 2- or 3-dimensional shape will grow more quickly and thus be more visually prominent than perhaps warranted (Huff 1954).

## Implied Relationships

Plotting two variables in the same visual field, such as with shared axes, can lead end-users to see relationships, associations, or correlations between those variables. Tyler Vigen runs a blog called "Spurious Correlations" (now a book) which draws attention to the absurdity of a range of unlikely statistical correlations. For instance, a chart taken from that blog shown in Figure 10.4 plots the number of people who drowned by falling into a pool versus the number of films that Nicolas Cage has appeared in. It visually indicates that the two lines have a good deal of correlation (there is also statistical correlation with Pearson $r = 0.67$). There's no explanation and no real reason for why these two variables should be related, yet such associations can easily be implied visually even when there is no logical rationale for a connection.

While such charts like Figure 10.4 are crafted to make light of how we often find meaningless relationships in big data, it is not uncommon to find published charts that imply an association between variables which

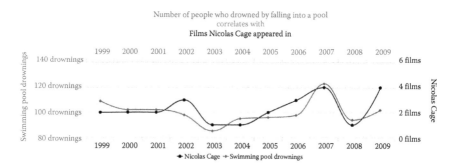

FIGURE 10.4   An absurd chart showing a spurious correlation that misleadingly connects two unrelated variables.

the human mind could easily interpret as causal. Take for instance the chart shown in Figure 10.5, which indicates the annual GDP growth in the United States according to whether there was a Democratic or Republican president in office. Simple looking at the top two bars in that graph visually shows that the economy *has* done better when there was a Democrat as president. But, thankfully, *The Economist* spends the next 2 minutes of the videographic from which this chart was excerpted explaining that in fact this association is not valid or causal and that research shows there are no factors that explain it other than lucky timing.

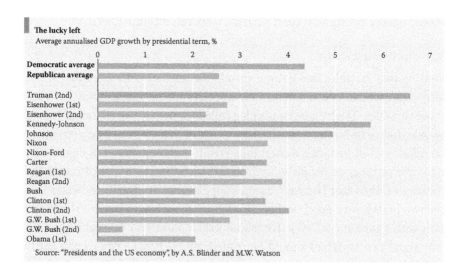

FIGURE 10.5   This chart suggests an association between the president in office and the robustness of the economy, but the rest of the videographic debunks this relationship by explaining the context around each president's term.

Another way in which relationships can be implied is through the sequencing and ordering of views within a narrative visualization (Hullman and Diakopoulos 2011; Hullman et al. 2013b). Temporal, spatial, or general to specific comparisons are often conveyed through subsequent views in interactive slideshow narratives or videographics; causal interpretations of such comparisons can result (Hullman et al. 2013b). Definitions of narrative revolve around the notion of a sequence of events that are causally related (Segel and Heer 2010). The human mind is prone to filling in causal explanations when we experience a story (Gottschall 2013), so it is particularly important to be aware of how readers may form interpretations of variable associations and if possible, to preempt common misinterpretations.

The overarching ethical point here is that data-driven storytellers must be careful that the variable relationships communicated in visualizations are not bogus or suggestive in a way that may be misleading. Whether those variable relationships are implied in the same visual space and chart, or whether they are implied through the sequencing and temporal juxtaposition of views, it is important to only convey those variable relationships (and their causal nature) when there is sufficient and reliable accompanying evidence, such as may come from additional research, context, or statistical analysis.

## Context and Annotation

The word "data" derives from a Latin term meaning "something given." It is therefore helpful to remember that data are not facts per se, but are "givens" from which a variety of interpretations can be derived. The meaning end-users impute from data is heavily moderated by how it is presented in relation to other data and to relevant context that sets the stage for interpretation. In visual storytelling, this is often accomplished with thorough labeling and textual explanation or annotation. Text provides that vital bit of context layered over the data that helps the audience come to a valid interpretation of what it really means. Annotations can emphasize, highlight, or prioritize particular pieces of data or interpretations, aiding the storytelling process by directing attention or preempting the user's curiosity on seeing a salient outlier, aberration, or trend (Hullman et al. 2013a).

One area that can be particularly fraught is in choosing labels for categories of people, as this inherently takes a stance on the existence and relevance of that category to the story, despite possible controversy over definitions or boundaries. This came up in a *New York Times* graphic in

2015 when a table representing lawmakers' support about the Iran nuclear deal included a column labeled "Jewish?"[*] The graphic was subsequently changed by removing the column, but not before provoking a reaction in which several viewers pointed out the irrelevance of the label "Jewish" in this political context.

Ethically, the goal in providing context and annotation is to provide context that is relevant, does not distract, and guides the reader towards the most plausible, logical, and faithful interpretation of the data. If a particular editorial interpretation of the data is conveyed via the context and annotations, then ideally this would also be indicated to the end-user. Care should be taken so that labels, legends, definitions, and other context are presented in a straightforward way that minimizes the potential for misinterpretation and which is complete enough for the viewer to fully comprehend the interpretation presented.

## Interactivity

Interactive visualizations provide capabilities for end-users to explore data and have agency as they move through the story presented. Designers can develop a dialogue with users rather than dictate a unitary interpretation that could convey a false sense of finality. Users can navigate, search, and filter the data according to their own interests, allowing for a greater degree of relevance and personalization. Some of the most consequential decisions that data-driven story designers make when creating interactive visual-izations relate to the *defaults*. Default views, default or suggested search terms, and default parameters such as slider values, all heavily bias end-users' interaction patterns. For some users, all they will see are the defaults. Ethically, it is important to choose defaults that are well-researched, based in evidence, and not arbitrarily chosen to provide an enticing or misleading view. *The New York Times* "Better to Rent or Buy?"[†] visual calculator does this well: the default inflation rate used by the calculator is 2.0%, which is right in the ballpark of values for the Consumer Price Index over the last few years.[‡] Besides defaults, other goal directions can also be embedded via interactivity such as by making suggestions for what the user should search for, or prompting the user to explore in particular ways rather than explore more freely (Hullman and Diakopoulos 2011).

---

[*] http://publiceditor.blogs.nytimes.com/2015/09/11/iran-deal-graphic-jewish-lawmakers-was-insensitive/.

[†] http://www.nytimes.com/interactive/2014/upshot/buy-rent-calculator.html.

[‡] http://data.bls.gov/timeseries/CUUR0000SA0L1E?output_view=pct_12mths.

TABLE 10.1   Ethical Considerations across Different Phases of Data-Driven Storytelling

| Data Acquisition | Data Transformation | Conveying and Connecting Insights |
|---|---|---|
| • Provenance<br>• Quantification | • Normalization<br>• Aggregation<br>• Algorithmic derivation<br>• Filtering<br>• Anonymization | • Visual mapping and representation<br>• Implied relationships<br>• Context and annotation<br>• Interactivity |

## SUMMARY

This chapter has enumerated a range of ethical factors (see Table 10.1) that are deserving of careful deliberation across the visual-data storytelling process. By considering tradeoffs in choices related to these various factors, the intent of this chapter is to help the data storyteller become more cognizant of their role and responsibility in guiding the interpretation of the audience. Whether in the data acquisition or transformation stages, when mapping data to charts, or connecting those charts into a sequence, ethics in this domain is predominantly about making sound editorial decisions that ensure the reception of accurate interpretations of the data. Of course, whether a story is interpreted accurately is also contingent on the data and visual literacy of the audience to begin with, and so ethical storytelling also attempts to understand the audience (see Chapter 9 entitled "Communicating Data to an Audience") and provide scaffolding of necessary literacies to ensure accurate interpretation. When errors do arise and there is a widespread unintended reception or interpretation that is untrue or misleading, the ethical storyteller will issue a correction, and update the story so that any misinterpretation can be avoided in the future. Striving to show the truth, being transparent, and engaging the community as an end is an ongoing endeavor in mindfulness, but one worth investing in to build a more-trustworthy craft of honest visual data storytelling.

## ACKNOWLEDGMENTS

This chapter benefitted greatly from the initial brainstorm and discussions on the topic with Gordon Bolduan, Marian Dörk, Kennedy Elliott, and Xaquín G.V. at the Dagstuhl seminar.

## REFERENCES

Bradshaw, Paul. 2015. "Data Journalism." In *Ethics for Digital Journalists: Emerging Best Practices*, edited by Lawrie Zion and David Craig. Routledge.
Cairo, Alberto. 2014a. "Ethical Infographics: In Data Visualization, Journalism Meets Engineering." *IRE Journal*, Spring 2014 Issue: 25–27.

————. 2014b. "Graphic Lies, Misleading Visuals." In *New Challenges for Data Design*, edited by Davic Bihanic, 103–16. Springer-Verlag, London.

————. 2017. "Moral Visualization." In *Information Design: Research and Practice*, edited by Sue Walker, Alison Black, Paul Luna, and Ole Lund. Routledge, Abingdon, UK.

Dick, Murray. 2016. "Developments in Infographics." In *The Routledge Companion to Digital Journalism Studies*, edited by Bob Franklin and Scott II Eldridge. Routledge, London.

Gottschall, Jonathan. 2013. *The Storytelling Animal: How Stories Make Us Human*. Mariner Books, New York.

Huff, Darrell. 1954. *How to Lie with Statistics*. Norton, New York.

Hullman, Jessica, and Nicholas Diakopoulos. 2011. "Visualization Rhetoric: Framing Effects in Narrative Visualization." *IEEE Transactions on Visualization and Computer Graphics* 17 (12): 2231–40.

Hullman, Jessica, Nicholas Diakopoulos, and Eytan Adar. 2013a. "Contextifier: Automatic Generation of Annotated Stock Visualizations." In *Proceedings of the Human Factors in Computing Systems (CHI)*.

Hullman, Jessica, Steven Drucker, Nathalie Henry Riche, Bongshin Lee, Danyel Fisher, and Eytan Adar. 2013b. "A Deeper Understanding of Sequence in Narrative Visualization." *IEEE Transactions on Visualization and Computer Graphics* 19 (12): 2406–15.

McBride, Kelly, and Tom Rosenstiel. 2013. "The New Ethics of Journalism : Principles for the 21st Century." In *The New Ethics of Journalism: Principles for the 21st Century*, edited by Kelly McBride and Tom Rosenstiel. CQ Press, Thousand Oaks, California.

Pandey, Anshul Vikram, Katharina Rall, Margaret L. Satterthwaite, Oded Nov, and Enrico Bertini. 2015. "How Deceptive Are Deceptive Visualizations?: An Empirical Analysis of Common Distortion Techniques." In *Proceedings of the Conference on Human Factors in Computing Systems*, 1469–78.

Powell, Wayne, and Mike Jempson. 2014. "More Accountability in the Digital Age? The Influence of New Technologies." In *Journalists and Media Accountability: An International Study of News People in the Digital Age*, edited by Susanne Fengler, Tobias Eberwein, Gianpietro Mazzoleni, Colin Porlezza, and Stephan Russ-Mohl 115–28.

Segel, Edward, and Jeff Heer. 2010. "Narrative Visualization: Telling Stories with Data." *IEEE Transactions on Visualization and Computer Graphics* 16 (6): 1139–48.

Stark, Jennifer, and Nicholas Diakopoulos. 2016. "Towards Editorial Transparency in Computational Journalism." In *Proceedings of Computation + Journalism Symposium*, Palo Alto, CA, September 30–October 1, 2016.

Stray, Jonathan. 2016. "Quantification." In *The Curious Journalist's Guide to Data*. Tow Center for Digital Journalism, New York.

Tufte, Edward. 2001. "Graphical Integrity." In *The Visual Display of Quantitative Information*, 2nd ed. Graphics Press, Cheshire, Connecticut.

Ward, Stephen J. A. 2015. *Radical Media Ethics: A Global Approach*. Wiley-Blackwell, Chichester, UK.

# CHAPTER 11

# Evaluating Data-Driven Stories and Storytelling Tools*

Fereshteh Amini

*University of Manitoba*

Matthew Brehmer

*Microsoft Research*

Gordon Bolduan

*Saarland University*

Christina Elmer

*Spiegel Online*

Benjamin Wiederkehr

*Interactive Things*

## CONTENTS

---

* Fereshteh Amini and Matthew Brehmer contributed equally to this chapter.

In this chapter, we review how data-driven stories and the tools used to produce them are evaluated. Evaluation is a far-reaching concept; among the topics we discuss in this chapter include the evaluation of a data-driven story in a newsroom context as well as the evaluation of novel storytelling tools and techniques in academic research settings. Our discussion spans a diverse set of goals, acknowledging the different perspectives of storytellers, publishers, readers, tool builders, and researchers. We review the possible criteria for assessing whether these goals are met, as well as evaluation methods and metrics that address these criteria. This chapter is intended to serve as a guide for those considering whether and how they should evaluate the stories they produce or the storytelling tools or techniques that they develop.

## INTRODUCTION

This chapter aspires to answer two questions drawing from a number of perspectives and disciplines. The first question pertains to how we should evaluate data-driven stories, while the second question pertains to how we should evaluate the tools and techniques developed to produce these stories.

Before we address these questions, however, it is also worth asking whether evaluation is even necessary or informative; what can we hope to learn from evaluating stories, tools, and techniques? To many, evaluation evokes laborious and methodical experiments and academic studies; in this chapter, we reveal how evaluation can assume many forms beyond academic experimentation, and that evaluation is not solely within the purview of researchers. We do realize that the different audiences for this chapter and book may have very different contexts and constraints, and we hope that our discussion of evaluation goals, criteria, methods, and metrics has some potential value to a wide array of storytellers, journalists, publishers, tool builders, and researchers.

The voices in this chapter include representation from the fields of journalism, design consulting, human-computer interaction, and information visualization, including both academic and nonacademic perspectives. Despite this multitude of voices and the variety of perspectives and context discussed within this chapter, our treatment of evaluation is surely not exhaustive or in any way prescriptive; our aim is rather to make readers aware of the range of evaluation criteria, methods, and metrics, as well as the connections and constraints affecting them. Another intent of this chapter is to provide links to relevant literature and to promote further cross-pollination between academic research and the applied practice of data-driven storytelling. Evaluation has been discussed in the human-computer interaction literature for decades [1], and it has more recently become a popular topic in the information visualization community [2,3]. Meanwhile, a number of journalists and publishers have questioned the efficacy of data-driven storytelling and standards by which their stories should be measured, and throughout this chapter we refer to a number of recent essays and blog posts by influential members of the journalism community that discuss this emerging topic.

### Outline

In the "Goals and Perspectives" section, we discuss the goals of various stakeholders related to data-driven storytelling. Then in the

FIGURE 11.1 Evaluation goals, criteria, methods, metrics, and constraints flagged for each perspective.

"Evaluation Criteria" section, we characterize the criteria by which these goals are met. The "Evaluation Methods" section follows, in which we enumerate a variety of methods for evaluating data-driven stories and tools for generating these stories. In the "Evaluation Metrics" section, we list objective metrics that can be recorded during the execution of the aforementioned methods that serve as proxies to the more-abstract evaluation criteria. Throughout the "Goals and Perspectives," "Evaluation Criteria," and "Evaluation Methods" sections, we provide links to related downstream concepts, such as from goals to criteria or from methods to specific metrics. The detailed list of all the concepts including links from different perspectives (i.e., audience, author, publisher, and tool builder) are depicted in Figure 11.1 to help the readers navigate through the chapter based on their take on the perspective for the evaluation of data-driven stories or storytelling tools. We also indicate the perspectives throughout using the following icons: 🖊 (author/storyteller), ★ (publisher), 👤 (audience), and 🔳 (tool builder). We conclude with a discussion in the "Challenges and Constraints" section on constraints affecting the choice of metrics and methods.

## GOALS AND PERSPECTIVES

We begin our discussion of evaluation by considering the perspectives of the various stakeholders involved in the data-driven storytelling process. By understanding and prioritizing the goals of these stakeholders, one can determine appropriate evaluation criteria, methods, and metrics. We consider the goals of the author or storyteller, the publisher, the storytelling tool builder, as well as the audience.

### Author/Storyteller Goals

Working at the intersection of several professions, authors of data-driven stories have to address different challenges and demands. The methodology behind their publications has to meet standards from the fields of statistics and computer science, whereas the visual outcome of their investigations will be judged mainly concerning its infographics and interaction design. However, most storytellers tend to see themselves primarily as journalists, and thus journalistic relevance is the most important benchmark for the evaluation of their stories. In this chapter, selected goals from the authors' perspective will be characterized and linked to useful evaluation methods.

### To Be the First to Break a Story

The news value of a data-driven story is closely related to its impact and the resulting fame accrued by its authoring journalists, programmers, and designers; "When great investigative work is paired with data journalism, remarkable outcomes bloom," writes Alexander Howard in his Tow Center for Digital Journalism report [4]. Data-based investigations often lead to newsworthy stories, often because these data sources have not been analyzed by journalists before or because these sources may be used to illuminate new issues.

Compared to conventional nondata-driven stories, data-driven stories and particularly those featuring data visualization have the advantage of being visually memorable and harder to copy by other media outlets. Furthermore, these stories often attract considerable attention on social media platforms. Readers who propagate data-driven stories may be perceived as smart or interesting, using impressive visuals to enhance their feeds.

### To Own a Story or Topic

In a digital media culture where stories circulate at enormous speed, journalists often have difficulties to "own" exclusive stories once they have published them. Data journalism techniques and interactive visualization can help these journalists to both find newsworthy topics and to keep those agenda-setting stories under control [5].

One early data-driven story featuring data visualization provides impressive evidence with regards to this strategy. In 2011, the German data journalism agency OpenDataCity and the news website *Zeit Online* published a map showing phone data and other freely available information of the politician Malte Spitz [6]; for the first time, this map revealed the far-reaching insights that such data sets can provide. The story was taken up globally, such as by *The New York Times*, who published an article on their front page with a direct reference to the German data-driven investigation and map [7].

### To Communicate, Inform, and Educate

The primary reason for telling data-driven stories is to communicate insights gathered from complex data and to inform the target audience about facts otherwise hidden in the data. The concept of data storytelling as a way of information delivery has been popularized by the leading media organizations such as *The New York Times* [8] and *The Guardian* [9]. Taking advantage of data-driven journalism supported by visualization, these organizations maintain their lead by making information more appealing for their audience. Data-driven stories can also be used for educational purposes. By creating stories, teachers can enhance lessons including facts supported by data as way of making abstract or conceptual content more understandable. Ma and Liao [10] have investigated how storytelling and visualization can make scientific findings more comprehensible and accessible to the general public. Scientists can tell a compelling story by focusing on important features and guiding the audience on how these features change over time or experimental conditions. The complexity of such data-driven lessons is highly dependent on the audience's level of visual literacy and how well the recipients of the data-driven story can interpret data visualizations supporting the narrative (see Chapter 9 entitled "Communicating Data to an Audience").

### To Indoctrinate and to Change Opinion

Various factors and reasons can result in incorrect knowledge or false beliefs about different topics such as social, economical, etc. Attempts to correct false beliefs are usually without success, but looking at real data evidence can be a way to correct persistently incorrect views of the world. An author of a data-driven story can take advantage of the power of narrative coupled with the right data visualization to enlighten an audience with insights that are difficult to see. This powerful combination can result in a data story that can influence and drive change. By starting a conversation on important issues surrounding a topic, the author can reach a wider audience and result in changing the collective opinion about the topic.

### To Persuade to Action or Change Behavior

Related to the use of data-driven stories as a way of communication, ultimately the communicator would like the audience to learn about the content presented through the story and be able to refer to it while making important decisions related to the subject presented in the story. For instance, a data-driven story presenting the data on the issue of global warming [11] aims at educating the audience about the issue by highlighting the evidence supporting the problem and hopefully resulting in the change in behavior and triggering action.

## To Facilitate Change in Policy and Governance

With the help of data-based investigations, reporters can scrutinize diverse relevant topics, be it government actions, political assertions, or social conditions. Thus, injustices can become apparent and discussed in the public dialogue. Following their self-image as watchdogs in the public interest, journalists use their sources and tools to unveil those scandals and bad developments. "Connections between powerful people or entities would go unrevealed, deaths caused by drug policies that would remain hidden, environmental policies that hurt our landscape would continue unabated. But each of the above was changed because of data that journalists have obtained, analyzed and provided to readers," says Cheryl Phillips, who worked as a Data Innovation Editor at *The Seattle Times* [12].

Or, as Simon Rogers, Data Editor at Google, points out, "data journalism can change the world" [13]. He follows this claim by giving two current examples published in early 2016: Fox News had used data graphics to fact check the Republican Party debate and thus changed its nature. And on a larger scale, ProPublica and the *Texas Tribune* had revealed that Houston, Texas is alarmingly unprepared for a big hurricane, likely to hit the city sooner or later. They did this by impressively combining interactive maps and sophisticated visuals within their story.

## To Be Validated, Recognized, and Acknowledged by Peers

Within their newsrooms, journalists working on data-driven stories featuring data visualization are often perceived as unicorns with special skills who apply methods to address topics that are hard to communicate. In contrast to this image, the community of these journalists can be characterized as extraordinarily cooperative and transparent. Within this community, there is a vibrant exchange of data sets, tools, and solutions used to handle common problems.

Projects that are decorated with international awards are of course celebrated among data journalists. Meanwhile, there are various prizes for data visualization, such as the Kantar Information is Beautiful Awards [14], the Malofiej Awards [15], and the Data Journalism Awards [16]. Moreover, visually driven stories are also successful in competitions lacking a distinct category for such work. For example, the 2016 Pulitzer Prizes awarded two projects that featured impressive interactive graphics, prompting Alberto Cairo, Knight Chair in Visual Journalism at the

University of Miami, to call for an infographics and data visualization category for the Pulitzer Prizes [17].

### To Appear as Being Aligned with Journalistic Values

By using rich interactive data-visualization tools and techniques within their reporting, journalists can demonstrate their support for distinct ideologies not referring to paradigms or worldviews, but to basic journalistic values. Above all, these values are closely related to those of investigative reporters, such as independence, a critical attitude, and a focus on injustices. Alexander Howard also describes this close link between data and watchdog journalism: "It's integral to a global strategy to support investigative journalism that holds the most powerful institutions and entities in the world accountable, from the wealthiest people on Earth, to those involved in organized crime, multinational corporations, legislators, and presidents" [4].

Besides, authors can also apply data visualization to make their work transparent by displaying single data points and providing both their source material and source code [18], either directly along with the primary article or via dedicated data-driven transparency channels such as Source [19]. Howard remarks that, "This ethos, where both the data and the code behind a story are open to the public for examination, is one that I heard cited frequently from the foremost practitioners of data journalism around the world" [4]. Thus, data journalists show parallels to civic hackers who also see the public good as a main purpose of their work and set high transparency standards for themselves and their peers.

### To Appear as Being Independent from Corporate Interests

As discussed in the previous section, journalists strongly value independence, especially for those working on investigative stories. Data journalists can follow this goal by analyzing source material and aggregating different data sets in addition to those provided by companies, interest groups, and think tanks. Journalists have to carefully check these sources for validity and reliability. Otherwise, confounders can corrupt and jeopardize the whole investigation.

Nicolas Kayser-Bril, a data journalist and cofounder of *Journalism++*, also emphasizes this tricky relationship: "Fluency with data will help journalists sharpen their critical sense when faced with numbers and will hopefully help them gain back some terrain in their exchanges with [public relations] departments" [12].

## Publisher Goals

Publishers and editors work behind the scenes and their names do not appear in the byline. Most of us only think of news publishers while watching movies like *Spotlight*, where American actor Liev Schreiber plays the editor-in-chief Marty Baron, or TV series like *The Newsroom*, where Jane Fonda acts as Leona Lansing, the CEO of the enterprise behind a fictional newsroom.

In this section, we describe the publisher perspective and what they might expect from data-driven storytelling. As we do not distinguish between CEO or editor-in-chief, we use the term publisher.

Both in fiction and in reality, publishers face economic pressure: whether it be due to falling circulation numbers or the increasing number of digital competitors, distribution channels, and changing format requirements. But it is not only cost-cutting that makes it necessary for publishers to define goals and derive criteria and metrics from them. Alex Howard, writing about the use of collaborative technology in enterprises, social media, and digital journalism reports: "I suspect that some newsrooms say they can't afford to hire newsroom developers when they really mean that their budget priorities lie elsewhere—priorities that are set by a senior leadership whose definition of journalism is pretty traditional and often excludes digital-native forms" [20].

All of these challenges have already been faced by the Berlin-based editorial room of the daily newspaper *Berliner Morgenpost* and its Editor-In-Chief Carsten Erdmann, who in 2013 hired the young journalist Julius Troger to design data-driven interactive pieces that independently analyze and aggregate different data sets. He also supported Troger with developers and graphic designers. "Creative output, strong usage numbers, and a growing expertise in data journalism enticed Funke Mediengruppe (one of Germany's larger publishers) to acquire [*Berliner Morgenpost*] in mid-2014" [21]. Hence, we will use their case to illustrate the aforementioned goals as follows.

### To Increase Page Views and Time Spent on Pages

It is often straightforward to transform the results of a data journalism project into basic graphics that can be published online. Readers appreciate them, and if the graphics are interactive, the readers may even stay longer on the website. Carsten Erdmann from *Berliner Morgenpost* discovered that very quickly. He reported: "The top ten most clicked stories in 2014 on morgenpost.de come from the Interactive team. Their stories go viral, reach new audiences, and generate reach. Now even advertisers ask for ads within those formats" [22].

In addition to page views, interactive graphics may also increase the amount of time readers spend viewing articles. Since data-based stories often provide insights into specific details, such as regional details, they tend to involve readers to a greater extent than conventional online news pieces. Thus, they can act as catalysts for media outlets who have to strengthen the loyalty of their readers to succeed in a highly competitive environment.

### To Increase Visibility on Social Media Channels

Data journalism and visual storytelling give publishers the opportunity to transform their results into formats streamlined for social media. Single infographics or short animated GIFs (Graphics Interchange Formats) can condense crucial details and convey the look and feel of an innovative news-reading experience. At the same time, they reflect well on the social media users that propagate them. Hence, publishers can push them via social media channels, earn likes, retweets, and other currencies of online attention, and thus increase the visibility of their brand.

### To Make a Business Sustainable

By practicing data-driven journalism, publishers have the opportunity to investigate stories that have not been previously covered. Alexander Howard quotes Scott Klein, Editor of News Applications at ProPublica, a nonprofit newsroom based in New York City: "There's no question that selling data is a rich opportunity for many newsrooms." Furthermore, visual storytelling based on such data sets is an interesting opportunity to generate profit. Paul Bascobert, Chief Operating Officer of the Bloomberg Media Group, explains: "It fits nicely within the trend towards more native, idea-driven and contextually relevant advertising" [23]. Hence, data-driven storytelling helps publishers sell their products and it attracts the attention of the advertising industry; both activities result in profit. Selling ads is still the lifeline of many publishers, regardless of whether the publisher is in the print or online business. Lucia Moses points out that publishers have two options: either they "[sell] price-depressing ads programmatically" or "[focus] on the long tail of lucrative, highly customized advertising (which presumably has a better shot at getting consumers' attention)" [23].

*To Increase Leadership*

The definition of leadership is the capacity of someone to lead a group. In the context of data-driven storytelling, it means the ability of a publisher to lead in a certain segment of journalism relative to its competitors. This ability is indicated by journalism awards, as the juries for these awards consist of experts within the profession.

As data journalism and data-driven storytelling are new forms, it is very attractive for every journalism award board to offer them as a category. Hence, a lot of awards exist for this field, making it easier for a publisher and an outstanding project to win a lot of them.

Julius Troger and his team at the daily newspaper *Berliner Morgenpost* proved that with "M29–The Bus Route of Contrasts," published in 2015 [24]. He explained the project in an interview: "The bus route cuts straight through Berlin. It starts in the villa districts in the west of the city, passes through the inner city areas, and ends in Berlin's trendiest districts. We collected and processed data regarding the neighborhood and local residents for every stop along the bus route. With the interactive application, we highlight the social differences of Berliners" [21].

In this way, the publisher got the attention of the most important journalist award programs in Germany:

> In 2015 it won the German reporter prize, awarded by Reporter Forum e.V., a network of German print journalists for outstanding reports in different media. Its official title is "German reporter prize, the prize of journalists for journalists."

> In 2016 it won the Nannen Prize (formerly Henri Nannen Prize) in the category "best Web reportage." This is a competitive prize, which recognizes the best journalistic work in print and online in the previous year. In the German media landscape, it is described as the "German Pulitzer Prize."

> In 2016 it was nominated for the Grimme Online Award, a spin-off of the famous "Grimme Award," one of the most prestigious awards for German television. "M29–The Bus Route of Contrasts" is competing in the Special category, which rewards innovative and outstanding concepts and examples of journalistic excellence.

On the international level, in 2015, the "M29" story was among the finalists of the Online Journalism Awards (OJAs) [25], honoring excellence in digital journalism around the world. The well-known Kantar Information is Beautiful Award [14] selected Julius Troger and his colleagues as Best Team in 2015.

### To Increase Engagement

"There are journalists doing great work to actively engage the audience in the storytelling process," says Amanda Zamora [26] who at the time was Senior Engagement Editor at ProPublica. Impact as a consequence of audience engagement is very important not only for these nonprofit newsrooms, but also for the traditional publishers, as their increasing hiring of so-called audience-engagement editors shows.

ProPublica's president, Richard Tofel defines engagement as "the intensity of reaction to a story, the degree to which it is shared, the extent to which it provokes action or interaction" [27]. Regarding online media consumption, engagement is often defined as the "time someone spends with content during a session or a given period of time" [28]. Earlier, we pointed out in the section entitled "To Increase Page Views and Time Spent on Pages" that interactive graphics may act as a catalyst for more activity on Web pages, apps, and social media platforms.

The British news outlet *The Guardian* provides a successful example. In 2015, the newspaper started the project "The Counted" [29], a data-driven approach to count the number of deaths caused by American police in that year and to cross-check official records. Within days, the project's Facebook page got about 15,000 likes. *The Guardian*'s investigative journalists used the tips and covered more than 1000 killings with detailed reports, resulting in an accessible database combining *The Guardian*'s reporting and verified crowdsourced information [30]. Mary Hamilton, the executive editor for audience at *The Guardian*, said: "It's become pretty clear that without building a living community of people who care about the issue around the journalism, the journalism would be much less successful."

### Data-Driven Storytelling Tool/Technique Developer Goals

Those who develop tools and techniques for data-driven storytelling may have several motivations. In some cases, the tool or technique developer is or has been an author of data-driven stories herself or himself and builds tools or techniques in order to improve or hasten the storytelling workflow. For instance, consider TimeLineCurator [31], a tool for authoring visual timeline stories; prior to developing this tool, the primary author had previously been generating timeline stories manually using illustration software, which was often a time-consuming and tedious process.

Though a tool or technique developer may be motivated by their own storytelling needs, developers often make their tools or techniques available, either publicly or within an organization, with the hope that their tool or technique can be adopted into the storytelling workflows or more generally be used by other people who are also telling data-driven stories.

Deploying a storytelling tool allows tool developers to study whether any adoption occurs. In cases where adoption does occur, developers can attempt to document case studies of people using the tool to author data-driven stories and to determine if the usage of the tool matches their expectations, or if the tool was appropriated for other purposes. Tool developers can also work with adopters to document usability issues, feature requests, and other requirements that can be addressed in future versions of the tool. Finally, there are cases in which the tool or technique developer also aspires to contribute to the research community, in which the deployment of a storytelling tool provides an opportunity to evaluate novel techniques or to document storytelling processes before and after a novel tool has been introduced.

## Audience Goals

In order to consider the goals of the audience for a data-driven story, we need to consider the contexts in which these stories appear and the relationship between the storytellers and their audience. In a recent article on the prospects for visualization research with respect to storytelling, Kosara and Mackinlay [32] distinguished contexts by whether the storyteller and audience are colocated, such as conference and boardroom presentations where the storyteller can pause to answer audience questions, and whether the story is consumed synchronously or asynchronously (e.g., a story communicated via a live televised newscast versus a story consumed via an interactive news graphic on a website).

In contexts where the audience and storyteller are colocated, such as at a conference, the audience is likely interested in the topic and may hope to learn something new, to stay informed, or hear new perspectives on the topic being presented. For example, Choe et al. [33] recently documented this form of data-driven storytelling at a Quantified Self conference in which the audience was united in an interest in self-tracking and were eager to hear the data-driven stories of other Quantified Selfers.

Like at a conference, another context in which the audience and storyteller are colocated is in the classroom. Ideally, students are also motivated to learn about a topic and develop their critical thinking skills.

Yet another context in which the audience and storyteller are colocated is in an organizational context, such as in a corporate management meeting or a public policy planning meeting. In these contexts, the audience may be motivated by a common need to make a policy decision.

For stories that are consumed asynchronously where the audience and storyteller are not colocated, we can refer to previous research documenting the consumption of data in casual contexts. For instance, previous research by Sprague and Tory [34] about the casual consumption of visualized data indicates that people are motivated to use visualization artifacts in these contexts by several intrinsic and extrinsic factors. Intrinsic factors include a desire to learn and understand the utility of the artifact, and a desire to be entertained. Extrinsic factors, on the other hand, include social pressure and a desire to avoid boredom. We can also refer to previous research that documents news reading behavior [35] and the motivation behind a person's decision to read a particular story; reading a newspaper or browsing online news often leads to no material gain and serves no "work" functions, however, it can induce moments of absorption. Though serendipitous discovery does occur in the context of news reading, Stephenson [35] found that people read most avidly what they already know about: a seemingly irrational activity that cannot be described as an explicit need to discover new information. Finally, reading news stories also serves the purpose of mutual socialization, giving people something to talk about in social settings.

Up to this point, the audience goals with respect to data-driven storytelling that we have discussed are no different from the goals of an audience for non data-driven storytelling: the audience is hoping to learn, to stay up-to-date, to be entertained, and to accumulate social fodder. However, it is important to recognize that there is also an audience for data-driven storytelling that cares greatly about the medium itself. Because of this audience, there are now conferences and awards dedicated to this genre of storytelling (e.g., the Kantar Information is Beautiful Awards [14]) as well as an interest in the technical solutions and design choices that lead to these data-driven stories (e.g., the Knight-Mozilla OpenNews Source project [19]). This audience will likely be interested in learning about the storyteller's process and the techniques that he or she used while researching and constructing his or her story.

## EVALUATION CRITERIA

Storytellers, publishers, tool builders, and audience members can assess whether their goals have been met according to a variety of criteria. In this section, we provide a nonexhaustive set of criteria by which data-driven stories and storytelling tools can be assessed. Whether these criteria are considered as part of an evaluation are subject to the values and goals of the people involved. Bear in mind that these criteria are constructs: their definitions vary across contexts and between people and are not directly measurable; metrics, on the other hand, which we discuss below in the section entitled "Evaluation Metrics" are directly measurable.

### Criteria for Evaluating Data-Driven Stories

To identify appropriate evaluation criteria with respect to different aspects of data-driven stories, we describe different criteria for authoring the data-driven story, consumption of the story by the audience, as well as the story impact postconsumption. Longing for impactful stories is inevitable if the end goal is to convey a message. As an example, the SketchStory study [36] focused on evaluating the story consumption (as well as the authoring process of the story), whereas the memorability study conducted in Borkin et al.'s study [37] focused on the story impact.

#### Comprehension

One of the main goals of a data-driven story is to have the audience understand the key points during its delivery and to remember them. Since it is not desirable for people to remember "wrong" information, knowing how well people understand the visualizations and collecting the insights they gained during the story consumption is a key aspect to assess. Visual analysis tools studied in the research community are typically targeted at experts such as data analysts or data scientists, who are knowledgeable in the data they want to study and are familiar with visualization, or are capable of investing significant amounts of time to learn complex visualization idioms. In contrast, data-driven stories may target a much broader audience, including people who may not have much literacy with respect to data and visualization. Thus, an important aspect to evaluate is the level of visual literacy required by the visualization. This topic is a growing interest in the visualization community and the VisLit Workshop [38] at the IEEE VIS 2014 conference allowed researchers to reflect upon visualization literacy measures. Furthermore, interactive data visualization is increasingly integrated into stories published online. It would be important to understand the readability of these data stories, which require people to understand not only visual representations but also how to interact with these representations.

### Memorability

Representing one of the most interesting criteria for evaluating the impact of data-driven stories is the question of whether or not the audience can recall (1) the content (e.g., different components [39], story units, and visualizations [37]), (2) the message(s) conveyed, and (3) the reasoning behind the story. Data-driven stories strive to deliver a message and therefore the ability to memorize and recall the message(s) is important. On the other hand, every element in a data-driven story serves that exact purpose (i.e., conveying the message); hence, by making these components memorable, individually, we can make sure that the main goal is achieved even more powerfully. Furthermore, the organization and sequence of these components to depict the broader image of the data-driven story must also be taken into consideration when measuring memorability to understand if the audience has followed the story.

### Engagement

With the explosion of information and stories that are freely available on the Web, it becomes a challenge to keep the audience's attention during the delivery of a story whether it is via a live presentation or an asynchronous sharing [32]. Therefore, it is important to evaluate the effectiveness of data-driven stories in terms of audience engagement during story delivery. There are multiple definitions of engagement, and engagement can be described as a continuum from low to high, from merely viewing to interacting to analyzing, synthesizing, and making decisions based upon these higher levels of engagement [40].

### Dissemination

An interesting criteria to evaluate data-driven stories is the extent to which a data-driven story is disseminated among an audience. This criteria is complex since there are several reasons as to why one might decide to share a story with peers. However, we can safely assume that if a data-driven story matches the goals of an audience, there is a higher chance of it being disseminated. For example, if someone is merely interested in being entertained, they would probably rate the stories they see based on the level of entertainment, and if they find the data-driven story highly entertaining, they might end up sharing it with friends.

### Increased Knowledge

While in many respects, data-driven stories, and in particular the use of visualization within a narrative, differs from the use of visualization for exploratory data analysis, we can draw a parallel between the two when it comes to the application of insight-based evaluation [41].

The purpose of a visualization is to facilitate insight extraction [42]. Data-driven stories aim to better communicate these insights to their target audience. Furthermore, understanding the key points or insight comprehension can be thought as two separate but closely coupled parts as described by Dove et al. [43]: the insight experience and the product of experience. The former refers to what psychologists describe as overcoming a mental block to achieve understanding. This occurs during the delivery of a data-driven story. Here, we like to highlight the latter, which represents the changed mental model resulting in new knowledge.

### Impact

If the intent of a data-driven story is to deliver a strong message, perhaps the most relevant impact can be seen as its echo in what journalists often call "real world." The effects can vary from changes in audience behavior, to influencing one's belief, all of which can ultimately result in an action.

Before the spread of stories on the Internet, there was no easy way of capturing data on the spread of a story and studying its "ripple effect." However, it is now possible to extract such data to generate hypotheses about possible connections between stories delivered via a narrative visualization and changes in the real world. The study of these relationships can also be conducted in a more-controlled way similar to Bond et al. [44] who performed a large-scale experiment that tested the influence of informational messages on voting behavior.

### Credibility and Trust

Different factors such as the source or quantity of data, the publisher, or the goal associated to the story may impact the credibility of the entire story independent of its design. For example, in the documentary film *An Inconvenient Truth* [45] directed by Davis Guggenheim, the former United States Vice President Al Gore had considerable credibility while educating citizens about global warming.

## Criteria for Evaluating Data-Driven Storytelling Tools

There are several criteria to consider when evaluating a data-driven storytelling tool: expressiveness, efficiency, usability, learnability, collaboration, and integration. Most tools do not excel with respect to all of these criteria; good tool designers attempt to balance these criteria, as there are typically trade-offs made when prioritizing one criteria over the others.

### Expressiveness

T    In the context of tools for storytelling, expressiveness relates to the range of possible stories that a storyteller could create using the tool, or the number of design choices that a storyteller could make. These design choices include the selection of visual representations and interactions, ranging from a constrained set of chart types and interactions such as mouse-over tooltips to totally custom visual encoding and novel interactions; for example, an author can create a story in Tableau [46] using a palette of prespecified chart types, while a storyteller can combine graphical primitive shapes using Lyra [47] to generate custom presentation-oriented charts. Another set of design choices pertains to annotation and how visual elements in a story can be annotated or highlighted with text, arrows, callouts, and other shapes. Additional design choices pertain to the presentation and sequencing of a story, ranging from a series of static charts to interactive linear slideshows and nonlinear semiexploratory stories; these presentation design choices will in turn constrain the set of possible narrative design patterns. Finally, there are low-level stylistic design choices including those pertaining to fonts, color palettes, and conventions for axes, legends, titles, and footnotes; full control over these choices will result in a distinctive look, perhaps aligning with the publisher's brand identity. For example, Mr. Chartmaker [48] will produce charts that adhere to *The New York Times'* style guide, whereas charts and stories produced by authoring tools such as Tableau [46] or Highcharts [49] tend to have a similar stylistic appearance.

### Efficiency

T    In the context of storytelling tools, efficiency relates to how quickly a storyteller or team of storytellers can produce a data-driven story using the storytelling tool, or how many stories can be produced using the tool in a set amount of time. This criteria is particularly important for journalists whose work must produce stories with short deadlines.

### Usability

T The usability of a storytelling tool refers to whether a person can accomplish their goal of producing a data-driven story using the tool. To determine whether a tool is useful, it is helpful to decompose the goal into a series of activities or tasks and determine if the tool supports these tasks. Once a tool is assumed to be useful, we can determine if the tool is usable by assessing the combinations of features within the interface of the tool that support the tasks: these features should be visible, identifiable, and accessible. Furthermore, as these tasks are performed, the result of using these features should be predictable, and appropriate feedback should be given such that the author can determine whether a task was accomplished. For a thorough discussion of usability, see Norman [1].

### Learnability

T The learnability of a storytelling tool is closely related to usability and the discoverability of features in a tool's interface. It is also related to expressiveness and efficiency: highly expressive tools may have more features, and thus an author will need more time to discover these features and learn how to appropriately use them; similarly, as one learns to use a tool, the authoring process will become more efficient. To increase the learnability of a tool, tool designers can provide documentation, interactive feature tours within the interface, and tutorial projects.

### Integration

T Another important criteria for a storytelling tool is how well it integrates into existing workflows, workplace contexts, and other tools. For instance, Mr. Chartmaker [48] integrates with *The New York Times'* content management system, while TimeLineCurator [31] integrates with TimelineJS [50], a popular slideshow application for timeline data. A related consideration is whether stories produced using the tool can be shared and embedded into different contexts.

### Extensibility

T Storytelling tools may not be highly expressive or well-integrated out of the box, but they may be extensible or modular in such a way that allows authors to customize or repurpose the existing features of the tool, or add new features that increases expressiveness or workflow integration.

### Collaboration

T Finally, collaboration in this context relates to whether multiple storytellers can use a tool to generate a story together. Additional considerations with respect to collaboration include whether storytellers can collaborate using multiple devices or gathered around a single device, and whether collaboration can take place synchronously or asynchronously. Collaboration support for data-driven storytelling tools can include features such as version control and the ability to comment on and discuss the story being produced.

## EVALUATION METHODS

In this section, we describe the methods by which the criteria defined above can be evaluated. We identify methods for evaluating data-driven stories and for evaluating storytelling tools and techniques.

Any evaluation method is associated with one or more metrics, types of data that are collected during the evaluation (we discuss metrics in detail in the following section). Some of these metrics are quantitative data while others are categorical or qualitative in nature, and as a result, evaluation methods tend to be described as being "quantitative" or "qualitative." Typically, an evaluation will involve the collection of multiple metrics via multiple methods and thus a mixed-methods evaluation comprised of both quantitative and qualitative methods and metrics is possible. It is through this combination of methods that we can triangulate on the evaluation criteria of interest.

There is extensive discussion of evaluation methods within the visualization research community [2,3] where methods include experimental studies of graphical perception, algorithmic benchmark tests, and contextual inquiries with people using visualization tools and techniques in practice. Not all of the methods discussed in this literature are appropriate for evaluating data-driven stories and the tools to create them; in this section, we will highlight several methods that are particularly appropriate for these stories and tools.

### Methods for Evaluating Data-Driven Stories

There are diverse sets of both quantitative and qualitative approaches for evaluating different criteria related to data-driven stories. In this section, we describe some of the evaluation approaches and instruments for the target criteria being evaluated.

### Collecting Performance Statistics

In order to evaluate data-driven stories, we can borrow some existing evaluation methods designed for exploratory data visualization systems, but such methods might not be perfectly adaptable to the nature of data-driven stories. For example, collecting statistics on meaningful interactions such as clicks, hovers, etc. can determine the level of audience engagement when viewing a data-driven story [51]. Similarly, it is possible to measure the time spent for the story consumption (or the number of times reviewing an earlier frame in data videos). Such time measurement, however, may be an indication of confusion when reading the data-driven story and not necessarily the indication of high a level of engagement. Therefore, adapting insight-based evaluations [41] to storytelling seems to be a more promising direction.

### Recall and Recognition Tests

Using insight-based evaluation techniques, we can employ recall and recognition tests to evaluate data-driven stories. For example, to evaluate the level of story comprehension or audience understanding of the content, one could attempt to capture the insights the audience gained shortly after the story consumption via recall and via recognition tests. Such tests could range from a simple questionnaire with questions about insights included in the data-driven story, to asking the audience to retell the entire story.

Similar tests can be used for measuring memorability of a data-driven story. By counting the number of components, story units, and visual elements a reader can remember ([37,52]), we can get an understanding on whether or not the audience remembers the intended key messages within the body of the data-driven story. While performing recall and recognition tests immediately after an audience is exposed to a narrative is indicative of retention, the later recall scores and results can be used to infer the long-lasting impact of the story.

### Questionnaires and Interviews

Self-reporting methods such as postviewing questionnaires (e.g., the Likert scale questionnaire), semi- and fully constructed interviews, as well as Product Reaction Cards [53] can be designed to evaluate data-driven stories. For instance, we could build a standard questionnaire to measure engagement, which will make it possible to compare different data-driven stories. Similarly, we can use questionnaires and audience comments to target whether or not a data-driven story has increased audience knowledge about a topic. Examples of such questions, derived from Product Reaction Cards [53], are listed below:

> What did you learn that you did not already know? In other words, describe new information/knowledge you gained.

> Did you learn something that contradicts what you already know about the topic [54]?

Answers to the above questions can also be correlated to different components within the data-driven story to learn about which components are most effective in terms of extracting knowledge.

Towards measuring the level of enjoyment and fun, researchers have used flow model-based questionnaires. For example, in the field of psychology, the well-known model of Csikszentmihalyi [55] has been employed to measure enjoyment. The application of a flow model to evaluate enjoyment in visualization including narrative visualization has recently been proposed [56].

Collecting audience ratings can also be a way of getting quick insights from the audience. For example, Amini et al. [57] compared the quality of data-driven videos created using DataClips with videos created using other tools by having an independent group of people view and rank the videos. Audience ratings can go beyond collecting numbers by breaking down the narrative into story units and asking the audience to not only rate each unit separately, but also to provide reasons as to why the particular rating was assigned. Doing so will provide deeper insight into the impact of the effectiveness of a data-driven story.

### Physiological Sensing

The audience consuming a data-driven story forms instantaneous emotional and physiological responses as a result of ongoing processing of the incoming information (i.e., data insights being communicated). Towards more objective measures for evaluating data-driven stories, we can leverage physiological sensing during the story consumption by collecting data on different physiological responses such as brain activity, blood pressure and heart rate, body and eye movement, pupil dilation, respiration, and skin conductance. Physiological sensing including neurophysiological and psychophysiological responses have been applied to marketing and traditional storytelling outlets to comprehend the cognitive and behavioral aspects of consumers [58]. However, this method has proven to be a challenging undertaking mainly due to the lack of appropriate sensing devices [59].

## Methods for Evaluating Data-Driven Storytelling Tools

To evaluate the expressiveness of a storytelling tool, one method is to generate a large number and a variety of stories based on a similarly wide variety of source data. A similar method involves recreating or approximating a wide variety of existing published stories.

### Usability Studies

To evaluate the usability and learnability of a storytelling tool, a simple usability study may be appropriate [60] in which participants naive to the tool are recruited and instructed to accomplish a specific goal or a series of specific tasks such as recreating or approximating an existing story or a series of presentation-oriented charts. As they attempt to complete these tasks, participants should be encouraged to think aloud while interacting with the interface of the tool. Think-aloud protocols are limited in that they do not capture automatic, nonconscious reactions to stimuli; as a result, a think-aloud protocol should be complemented with a retrospective interview or survey following the completion of the tasks. Usability study sessions should also be video recorded, and if possible, the authoring tool should be instrumented such that an interaction log is generated as a person interacts with the tool. These additional sources of data, along with the think-aloud transcript result in qualitative insights about the tool that can help the tool evaluators to assess whether the tool's functionality is visible, accessible, and used as they were anticipated by the designers, as well as whether the use of features produce the results by the participants. These usability studies can also help to determine if the tool requires additional functionality, or if existing functionality is unnecessary or redundant.

An example of an informal usability study is the evaluation of iVis-Designer [61], an interactive visualization tool. The authors recruited eight people and demonstrated the features of the tool via a short tutorial video. The authors encouraged the participants to follow along with the video, which featured three charts that could be authored using the tool. Following the recreation of these charts, the participants were free to generate their own designs. Throughout both stages, participants followed a think-aloud protocol.

### First-Use Studies

[T] A variant of the usability study is the first-use study; Satyanarayan and Heer conducted such a first-use study as a means to evaluate whether Ellipsis [62], a narrative visualization storytelling tool, fit the needs and workflows of journalists. After demonstrating the features of Ellipsis to eight journalists, Satyanarayan and Heer asked the journalists to author a short story with a particular set of charts based on a familiar story published by *The New York Times*. The same authors conducted a similar first-use study with Lyra [47], an interactive presentation-oriented visualization design environment. In that study, they recruited 15 people including a mix of journalists, designers, analysts, and students. Following a brief tutorial, the participants were asked to recreate three charts of increasing complexity while thinking aloud.

### A/B Testing

[T] Another method for evaluating a data-driven storytelling tool is a comparative A/B study, either comparing different versions of a tool containing alternative designs, or comparing a prototype tool to an existing tool. In the latter case, evaluators should ensure a fair comparison: the two tools should accomplish the same goal and have comparable functionality, and comparable or identical source data should be supplied to participants for each version. The purpose of comparing different tools would be to compare them in terms of efficiency, usability, and learnability. An example of such a comparative evaluation was a comparison of TimeLineCurator's timeline story producing process [31] to the timeline story producing process of TimelineJS [50] in which participants authored the same story with both tools. Regardless of whether the comparative evaluation is comparing different tools or versions of the same tool with alternative feature designs, the presentation order of the tools or tool versions should be counterbalanced across participants to reduce the effects of learning and fatigue. Evaluators should consider recording the time that participants take to complete a task or series of tasks along with any qualitative feedback they provide via a think-aloud protocol or a retrospective semistructured interview. Since the resulting stories may differ substantially and their content will be subject to each participant's authorial judgment, it is difficult to gauge what accuracy or an error means in this context. The resulting stories can, however, be evaluated qualitatively with a focus on their content and complexity.

### Case Studies

Finally, data-driven storytelling tools can and should be evaluated in the wild: with real storytellers telling stories with data that is personally or professionally relevant to them. One approach is to recruit journalists or data analysts who need to communicate their findings as part of their work. For example, the designers of iVisDesigner [61] recruited a data analyst who wanted to communicate findings from his analysis of social media data; they were able to ingest his data and they reported on a case study of this analyst using iVisDesigner to create presentation-oriented charts. Another approach is to deploy and promote a tool online, monitor its usage information, and solicit feedback from the community of early adopters, such as in the case of TimeLineCurator [31]. This approach can provide opportunities to report on case studies of people using the tool in the wild.

## EVALUATION METRICS

In this section, we comment on specific evaluation metrics, or data collected via the evaluation methods described in the previous section, being directly measurable proxies for the evaluation criteria as discussed in the section entitled "Evaluation Criteria."

### Quantitative Metrics for Stories

Nowadays, advanced analytics tools can produce reams of quantitative data such as the number of views, unique viewers, likes (i.e., ratings), shares (i.e., reposts), references made, and more. When quantitative metrics are analyzed to ascertain the impact of a story, many confounds have to be taken into account especially when it comes to comparisons between two stories. This is due to the fact that the personal engagement level is low. In this sense, the recorded information is prone to numerous biases (multiple ratings by the same person, buzz word, etc.). Also, these ratings merge many criteria and only give a global and personal assessment. To produce reliable conclusions, one would need an experimental design where both underlie the absolutely same conditions. But in reality, these can differ in many ways. For example, the specific moment in time when a piece is published can be crucial concerning its audience and performance, as well as its positioning on the website or other distribution channels used.

In addition to the metric of time, supportive measures on social media platforms can also make a difference. This is crucial to keep in mind if you analyze metrics originating from the measurement, the collection, and reporting of Internet data (Web analytics). First and foremost, they are used to get a better understanding of users consuming content online on a single web page. Nevertheless, they can provide valuable hints on how the audience reacts to your published data-driven stories.

*Page Views*

Out of a technical perspective, a page view is a request to load a piece of content on a single Web page, also described as a page impression. This request can be initiated by several activities, e.g., clicking on a link or just refreshing a page. In this context, the terms hit and click have a more general meaning as they do not specify the location of the request. Together they are the most close-grained metric for Web analytics.

*Time Spent on a Page*

Of course, the time that users spend on a Web page might indicate how interesting its content is. This is measured by engaged time and page views per visit. Engaged time is the amount of time users do something on the page like scrolling, writing, and watching [28]. Page views per visit or page views per session, in contrast, describe a collection of hits during a predefined amount of time (30 minutes in the most cases).

*Number of Users*

A user is a person indicated by a pseudonymous identification string while accessing a Web page. Hence, the number of users is an estimation of how many different persons are using the Web page. Social media platforms such as Facebook and Twitter provide a multitude of metrics. The most important and most basic include reactions/comments and reshares, as explained next.

*Number of Likes, Comments, and Replies*

A like generally means a person appreciates a certain post. They indicate that by clicking on a predefined button. Similarly, we have a number of persons reacting on your post by replying or commenting on it via a short text message.

### Number of Shares/Retweets

Another metric is the number of people sharing a specific post of another person or institution. Metrics, like followers and friends, are also very important social media analytics. But as they refer more to the accounts of persons and institutions, they are not important in the context of stories published via these channels. Hence, we have not itemized these here.

### Awards/Recognition

A completely different metric is public awards. For data journalism and data-driven storytelling, a lot of them exist on a national and international level. Hence, the number of awards a story gets can serve as a quantitative metric, too. For details, see the section entitled "To Increase Leadership."

### Physiological Responses

Physiological responses and signals collected during the consumption of a data-driven story can be used to understand the processing of information being presented. For example, collecting statistics about ocular movements of the audience while reading or viewing a story has been used to compare stories generated using different strategies (i.e., infographics vs. newspaper articles) [63]. With the advancements of sensing devices, other neurophysiological and psychophysiological responses can also be applied to human-computer interaction (HCI) related research including the evaluation of data-driven stories.

Relying solely on this metric may not provide complete data; however, it would be still possible to gain initial understanding of the audience's affective state. Furthermore, when used in conjunction with data gathered through other methods (e.g., audience comments, questionnaires), physiological responses may shed more light on how much the audience was engaged with a data-driven story.

## Categorical Metrics for Stories

To gain more credible insights when using quantitative metrics, the users can be clustered following certain categories. By comparing these groups, the evaluation of one single piece can be further intensified, leading to sophisticated findings that can also be used in comparisons between several stories.

*Traffic Sources*

T ⋆ Readers can reach stories from various directions, resulting in different usage settings and levels of involvement. Thus, clustering the traffic based on its sources can help newsrooms to evaluate different distribution channels they use and the visibility of their output for search algorithms. Chartbeat, one of the most common metrics vendors, distinguishes between five categories of traffic sources: direct, social, external, internal, and search [64].

Direct traffic includes readers who typed in a specific uniform resource locator (URL) or used a bookmark, hence, to reach the website consciously. In contrast, users who arrive there after a more general search request are less likely to be loyal users already. External traffic includes readers coming from a link on an external website, not via search engines or social media platforms. By clustering the traffic in this way and comparing quantitative metrics like reading time, descent, or conversion rates, newsrooms can draw conclusions about the behavior, motivation, and accessibility of different reader groups. This can lead to various consequences and adjustments in publication manners, especially dependent on the medium's strategy and target audience.

*Devices*

T ⋆ The user experience of a data-driven story is clearly determined by the device's type, screen size, and other features, especially when it comes to complex interactive data visualizations. Concerning their device, readers are usually clustered into three groups: desktop, mobile, and tablet visitors [64]. These platforms can be further segregated, since they all offer more than one way to access a story, for example, via a browser or a news app. The share of the different devices is closely related to varying usage situations and patterns, repeating themselves on a daily or weekly basis.

In times when more and more users are consuming news on mobile devices, media outlets have to adjust their storytelling and data-visualization techniques. As Len DeGroot, the director of data visualization for *The Los Angeles Times*, points out, mobile should not be regarded as a simplified version of desktop: "These devices are there, they have these capabilities built into them, so how do we use it in our storytelling?" [65]. Following the trend towards a mobile-dominated usage, newsrooms are yet going over to designing their articles and applications mobile-first.

By evaluating visually driven stories using quantitative metrics, the comparison of usage situations and devices can lead to valuable insights. If the time spent on a page with an interactive data visualization strongly differs between mobile and desktop users, newsrooms may think of a more-advanced adjustment of concepts, features, and designs. Furthermore, they can use metrics to evaluate both the interaction intensity of their readers and measures to foster their engagement.

### Outgoing Traffic Destination

To increase their impact and advertising revenues, newsrooms not only have to gain new visits, but also attract new loyal readers who are regularly spending time on a website and are interacting with more than one news piece. Thus, analyzing the destination of a story's outgoing traffic can produce valuable insights. Several quantitative categories can be used to determine what happens after a visually driven article has been read and explored by a user.

At worst, this story is the first and last page visited before the reader leaves the website. The Web Analytics Association (WAA) names this a "single page view visit." The percentage of such users calculated for a specific page is called the bounce rate [64]. Their detailed behavior on a page can yield helpful clues about the hurdles that are hard to overcome, which is of special value concerning interactive data visualizations that often combine both guidance and exploration functionalities.

To prevent a story from becoming a so-called exit page, newsrooms can highlight links to other related pieces on their website with similar contents or display formats. Thus, the recirculation rate can be increased, including the audience who has engaged with a particular piece of content and proceeds to engage with another piece of content.

### Qualitative Metrics for Stories

Collecting qualitative metrics about a data-driven story can be used as a way of directing answering questions about target criteria or explaining and confirming insights generated through quantitative instruments and metrics.

For example, we can analyze the audience response to a story by collecting their comments, especially when a story is published via a platform with an existing comment space. One foci of interest when analyzing comments on data-driven stories is whether the audience discusses the context around the story or specific data featured in the story [66].

A related foci of interest when analyzing audience comments in response to a data-driven story is whether the audience is commenting on and discussing the particular elements or trends in the data highlighted by the author or if the audience is discussing other elements or trends not discussed by the author, as well as whether the audience is acknowledging, supporting, or refuting the data-driven conclusions of the author [67].

Another metric of interest is a change in reader opinion. One way to collect this data is to gather the opinions of an audience about a subject or topic prior to exposing them to a story. After consuming the story, the audience can be polled again to assess whether any change of opinion has taken place and whether this change of opinion can be attributed to specific data points or trends featured in the story.

A final metric worth mentioning is any number of actions or reforms taken by organizations, governments, or politicians in response to a story; an example would be new regulations on the use of offshore tax havens in response to data-driven stories resulting from the Panama Papers leak in early 2016. Such actions or reforms may be difficult to directly associate with a particular story, especially when a series of stories about a topic results in public outcry and discussion, leading to increased pressure on decision makers who ultimately take action.

## CHALLENGES AND CONSTRAINTS

It might be incomplete to itemize and explain methods and metrics for the evaluation of data-driven stories without pointing out the related constraints and confounds. In this section, we comment on some of the challenges editorial departments and data-driven storytellers have to cope with in real life.

Perhaps the first challenge we face when aiming to evaluate a data-driven story is narrowing down the criteria most appropriate for assessing goals.

Having a vision with clear goals may also be a challenge if the researcher does not necessarily have a perspective to focus on when assessing the data-driven story. For instance, a publisher with the goal to attract more audiences to a published story can be compared to an independent researcher who needs to justify why the perspective taken is valid and worth investigating.

Once the right criteria is decided on, there are several metrics and instruments to employ for the evaluation as we have commented on in this chapter. However, the task of adapting the right methods can prove to be difficult due to several factors such as the story context, the complexity of data and insights, the story medium, etc. Furthermore, a variety of constraints and confounds can affect this process and must be taken into consideration. We conclude this chapter by highlighting some of these constraints.

## Human Resources and Expertise

T ★ Specialists are expensive. As editorial rooms have to fight cost savings, they have no budget to pay data analysts and or even scientists who can guide the evaluation process. Even worse, often there are not enough editors in the newsroom to do the evaluation, for example by doing simple A/B tests. The small budget restricts the use of professional tools like Google Analytics, Metrics, and Adobe. Some of these services are very expensive.

## Time/Deadlines

★ If you work in a journalistic domain, most of the time you have to produce data-driven stories in a predefined time and to meet the deadline. Exceeding the deadline makes the produced story worthless. One consequence of this pressure is that you concentrate only on the necessary parts and neglect aspects which are time consuming but also increase the quality of the story.

## Budget

★ Time and money have always been precious in editorial departments and newswires. Since the media crisis this has intensified. Hence, it can be a real challenge to organize enough budget to employ a team of experts to evaluate the quality of a data-driven story.

## External Validity

T ★ It is very difficult to compare their own evaluation results with those of other websites as the underlying metrics are not selective enough and the conditions are very different. Additional to that, technology companies like Google and Facebook add a fuzziness to the evaluation of success because there is a lack of specific knowledge on how their algorithmic curation may affect results.

## Conclusion

In this chapter, we discussed the topic of how data-driven stories and the tools used to produce them are evaluated. Evaluation is a wide-reaching concept, and the term evokes different meanings in different domains:

evaluation of a data-driven story in a newsroom will be very different from the evaluation of a novel storytelling technique in an academic research setting. For this reason, we classified the diverse set of goals of storytellers, publishers, readers, tool builders, and researchers. Given these goals, we then enumerated possible criteria for assessing whether these goals are met, as well as evaluation methods and metrics that approximate these criteria. It is our intent that this chapter serves as a rough guide for those considering whether and how they should evaluate the stories they produce or the storytelling tools or techniques that they develop.

## REFERENCES

1. D. A. Norman. *The Psychology of Everyday Things*. Basic Books, New York, 1988.
2. S. Carpendale. Evaluating information visualizations. In A. Kerren, J. T. Stasko, J.-D. Fekete, and C. North, editors, *Information Visualization: Human-Centered Issues and Perspectives*, chapter 2, pp. 19–45. Springer, New York, 2008. http://dx.doi.org/10.1007/978-3-540-70956-5_2.
3. H. Lam, E. Bertini, P. Isenberg, C. Plaisant, and S. Carpendale. Empirical studies in information visualization: Seven scenarios. *IEEE Transactions on Visualization and Computer Graphics (TVCG)*, 18(9):1520–1536, 2012. http://dx.doi.org/10.1109/TVCG.2011.279.
4. A. Howard. The art and science of data-driven journalism, December 2014. http://towcenter.org/research/the-art-and-science-of-data-driven-journalism/.
5. London Press Club. The numbers game debate: "Data journalism gets exclusives", May 2016. http://londonpressclub.co.uk/thenumbers-game-debate-data-journalism-gets-exclusives/.
6. Zeit Online and OpenDataCity. Tell-all telephone, March 2011. http://www.zeit.de/datenschutz/malte-spitz-data-retention.
7. N. Cohen. It's tracking your every move and you may not even know, March 2011. http://www.nytimes.com/2011/03/26/business/ media/26privacy.html.
8. The New York Times. http://www.nytimes.com/.
9. The Guardian Datablog. http://www.theguardian.com/data/.
10. K.-L. Ma, I. Liao, J. Frazier, H. Hauser, and H.-N. Kostis. Scientific storytelling using visualization. *IEEE Computer Graphics and Applications*, 32(1):12–19, 2012.
11. Jet Propulsion Laboratory, National Aeronautics and Space Administration. Global climate change: Evidence, 2016. http://climate.nasa.gov/evidence.
12. J. Gray, L. Bounegru, and L. Chambers, editors. *The Data Journalism Handbook*, O'Reilly Media, Boston, 2011. http://datajournalismhandbook.org/1. 0/en/index.html.

13. S. Rogers. Data journalism matters more now than ever before, March 2016. https://simonrogers.net/2016/03/07/data-journalism-matters-more-now-than-ever-before/.

14. Kantar. Information is Beautiful Awards showcase, 2016. http://informationisbeautifulawards.com/showcase.

15. Society for News Design-Europe. Malofiej Infographics Word Summit & SDT, 2016. http://www.malofiejgraphics.com/.

16. Global Editors Network. Data Journalism Awards, 2016. http://www.globaleditorsnetwork.org/programmes/data-journalism-awards/.

17. A. Cairo. It's time for a Pulitzer Prize for infographics and data visualization, April 2016. http://www.thefunctionalart.com/2016/04/it-is-time-for-pulitzer-prize-for.html.

18. J. Stark and N. Diakopoulos. Towards editorial transparency in computational journalism. In *Proceedings of the Symposium on Computation + Journalism*, 2016. https://goo.gl/xLLAS7.

19. Knight-Mozilla OpenNews. Source, 2016. https://source.opennews.org/.

20. A. Howard. Publishers can afford data journalism, says ProPublica's Scott Klein, April 2014. http://towcenter.org/publishers-can-afford-data-journalism-says-propublicas-scott-klein/.

21. M. Lorenz. The story of a transformation, in three years, February 2015. http://datadrivenjournalism.net/news_and_analysis/the_story_of_a_transformation_in_three_years.

22. J. Twiehau. Grenzenloses arbeiten. *Medium Magazin*, (07/2015), 2015. http://www.mediummagazin.de/inhalt/medium-magazin-072015/.

23. L. Moses. Is there an ad model for explainer journalism? April 2014. http://digiday.com/publishers/ad-model-explainer-journalism/.

24. J. Troger. M29 the bus route of contrasts, January 2015. http://interaktiv.morgenpost.de/m29/.

25. Online News Association. Online Journalism Awards, 2016. http://journalists.org/awards/.

26. A. Zamarro. A closer look: I'm not (just) your paperboy why more newsrooms should embrace crowd-powered journalism, September 2015. https://www.propublica.org/article/a-closer-look-im-not-just-your-paperboy.

27. L. Green-Barber. How can journalists measure the impact of their work? Notes toward a model measurement, March 2014. http://goo.gl/ 1Pcgwk.

28. F. Cherubini and R. K. Nielson. Editorial analytics: How news media are developing and using audience data and metrics, February 2016. http://papers.ssrn.com/sol3/papcrs.cfm?abstract_id=2739328.

29. Guardian staff. The counted: Tracking people killed by police in the United States, 2015. https://www.theguardian.com/us-news/series/counted-us-police-killings.

30. Guardian staff. Inside the counted: How Guardian US has tracked police killings nationwide, April 2016. https://www.theguardian.com/membership/2016/apr/11/inside-the-counted-guardian-us-police-killings.

31. J. Fulda, M. Brehmer, and T. Munzner. TimeLineCurator: Interactive authoring of visual timelines from unstructured text. *IEEE Transactions on Visualization and Computer Graphics (Proceedings of VAST 2015)*, 22(1):300–309, 2016. http://dx.doi.org/10.1109/TVCG.2015. 2467531.

32. R. Kosara and J. D. Mackinlay. Storytelling: The next step for visualization. *IEEE Computer*, 46(5):44–50, 2013. http://doi.ieeecomputersociety. org/10.1109/MC.2013.36.

33. E. K. Choe, B. Lee, and M. Schraefel. Characterizing visualization insights from quantified selfers' personal data presentations. *IEEE Transactions on Computer Graphics and Applications (CG&A)*, 4(35):28–37, 2015. http:// dx.doi.org/10.1109/MCG.2015.51.

34. D. Sprague and M. Tory. Exploring how and why people use visualizations in casual contexts: Modeling user goals and regulated motivations. *Information Visualization*, 11(2):106–123, 2012. http://dx.doi. org/10.1177/1473871611433710.

35. W. Stephenson. *The Play Theory of Mass Communication*. University of Chicago Press, Chicago, 1967.

36. B. Lee, R. H. Kazi, and G. Smith. SketchStory: Telling more engaging stories with data through freeform sketching. *IEEE Transactions on Visualization and Computer Graphics (Proceedings of InfoVis)*, 19(12):2416–2425, 2013. http://dx.doi.org/10.1109/ TVCG.2013.191.

37. M. A. Borkin, A. A. Vo, Z. Bylinskii, P. Isola, S. Sunkavalli, A. Oliva, and H. Pfister. What makes a visualization memorable? *IEEE Transactions on Visualization and Computer Graphics (Proceedings of InfoVis)*, 19(12):2306–2315, 2013. http://dx.doi.org/10.1109/TVCG.2013.234.

38. S.-H. Kim, J. Boy S. Lee J. S. Yi, and N. Elmqvist. Towards an open visualization literacy testing platform, 2014. http://ieeevis.org/year/2014/info/ overview-amp-topics/accepted-workshops.

39. S. Bateman, R. L. Mandryk, C. Gutwin, A. Genest, D. McDine, and C. Brooks. Useful junk?: The effects of visual embellishment on comprehension and memorability of charts. In *Proceedings of the ACM Conference on Human Factors in Computing Systems (CHI)*, pp. 2573–2582, 2010. https:// doi.org/10.1145/1753326.1753716.

40. N. Mahyar, S.-H. Kim, and B. C. Kwon. Towards a taxonomy for evaluating user engagement in information visualization. In *Proceedings of the IEEE VIS 2015 workshop Personal Visualization: Exploring Data in Everyday Life*, 2015. http://www.vis4me.com/personalvis15/papers/mahyar.pdf.

41. P. Saraiya, C. North, V. Lam, and K. Duca. An insight-based longitudinal study of visual analytics. *IEEE Transactions on Visualization and Computer Graphics (TVCG)*, 12(6):1511–1522, 2006. http://dx.doi.org/10.1109/ TVCG.2006.85.

42. S. K. Card, J. D. Mackinlay, and B. Shneiderman. *Readings in Information Visualization: Using Vision to Think*. Morgan Kaufmann, Burlington, MA,1999.

43. G. Dove and S. Jones. Narrative visualization: Sharing insights into complex data. In *Conference Proceedings of Interfaces and Human Computer Interaction (IHCI)*, 2012. http://openaccess.city.ac.uk/1134/.

44. R. M. Bond, C. J. Fariss, J. J. Jones, A. D. Kramer, C. Marlow, J. E. Settle, and J. H. Fowler. A 61-million-person experiment in social influence and political mobilization. *Nature*, 489(7415):295–298, 2012. http://dx.doi. org/10.1038/nature11421.

45. D. Guggenheim (Director). *An Inconvenient Truth*. Participant Media. 2006, www.participantmedia.com.

46. Tableau Software. Tableau, 2016. http://tableau.com.

47. A. Satyanarayan and J. Heer. Lyra: An interactive visualization design environment. *Computer Graphics Forum (Proceedings of EuroVis)*, 33(3), 2014. http://dx.doi.org/10.1111/cgf.12391.

48. G. Aisch. Seven features you'll want in your next charting tool. Presented at NICAR, July 2015. http://goo.gl/cywIqO.

49. A. S. Highsoft. Highcharts, 2016. http://www.highcharts.com.

50. Northwestern University Knight Lab. TimelineJS, 2013. http://timeline. knightlab.com/.

51. J. Boy, F. Detienne, and J.-D. Fekete. Storytelling in information visualizations: Does it engage users to explore data? In *Proceedings of the ACM Conference on Human Factors in Computing Systems (CHI)*, pp. 1449–1458, 2015. https://doi.org/10.1145/2702123.2702452.

52. M. A. Borkin, Z. Bylinskii, N. W. Kim, C. M. Bainbridge, C. S. Yeh, D. Borkin, H. Pfister, and A. Oliva. Beyond memorability: Visualization recognition and recall. *IEEE Transactions on Visualization and Computer Graphics (Proceedings of InfoVis 2015)*, 22(1):519–528, 2016. http://massvis. mit.edu/; http://dx.doi.org/10.1109/ TVCG.2015.2467732.

53. T. Mercun. Evaluation of information visualization techniques: Analysing user experience with reaction cards. In *Proceedings of the ACM Workshop on Beyond Time and Errors: Novel Evaluation Methods for Visualization (BELIV)*, pp. 103–109, 2014. https://doi. org/10.1145/2669557.2669565.

54. D. Badawood and J. Wood. The effect of information visualization delivery on narrative construction and development. In *Proceedings EuroVis Short Papers*, 2014. http://dx.doi.org/10.2312/eurovisshort.20141148.

55. M. Csikszentmihalyi. *Flow: The Psychology of Optimal Experience*. New York, 1990.

56. B. Saket, A. Endert, and J. Stasko. Beyond usability and performance: A review of user experience focused evaluations in visualization. In *Proceedings of the Beyond Time and Errors on Novel Evaluation Methods for Visualization*, pp. 133–142. ACM, 2016.

57. F. Amini, N. H. Riche, B. Lee, A. Monroy-Hernandez, and P. Irani. Authoring data-driven videos with dataclips. *IEEE Transactions on Visualization and Computer Graphics (Proceedings of InfoVis 2016)*, 23(1):501–510, 2017. https://doi.org/10.1109/TVCG.2016. 2598647.

58. T. Yang, D.-Y. Lee, Y. Kwak, J. Choi, C. Kim, and S.-P. Kim. Evaluation of TV commercials using neurophysiological responses. *Journal of Physiological Anthropology*, 34(1):19, 2015.

59. C. Peter, E. Ebert, and H. Beikirch. Physiological sensing for affective computing. In *Affective Information Processing*, pp. 293–310. Springer-Verlag, London, 2009.

60. J. Nielsen. Why you only need to test with 5 users, March 19, 2000. http://goo.gl/6Huppn.

61. D. Ren, T. Hollerer, and X. Yuan. iVisDesigner: Expressive interactive design of information visualizations. *IEEE Transactions on Visualization and Computer Graphics (Proceedings of InfoVis)*, 20(12):2092–2101, 2014. http://dx.doi.org/10.1109/TVCG.2014.2346291.

62. A. Satyanarayan and J. Heer. Authoring narrative visualizations with Ellipsis. *Computer Graphics Forum (Proceedings of EuroVis)*, 33(3):361–370, 2014. http://dx.doi.org/10.1111/cgf.12392.

63. F. De Simone, F. Protti, and R. Presta. Evaluating data storytelling strategies: A case study on urban changes. In *COGNITIVE 2014, The Sixth International Conference on Advanced Cognitive Technologies and Applications*, pp. 250–255. Citeseer, 2014.

64. C. Petre. The traffic factories: Metrics at Chartbeat, Gawker Media, and The New York Times, May 2015. http://towcenter.org/research/traffic-factories/.

65. S. Wang. Small screens, full art, can't lose: Despite their size, phones open up new opportunities for interactives, March 2016. http://goo.gl/OQywhW.

66. W. Willett, J. Heer, J. Hellerstein, and M. Agrawala. Commentspace: Structured support for collaborative visual analysis. In *Proceedings of the ACM Conference on Human Factors in Computing Systems (CHI)*, pp. 3131–3140, 2011. http://doi.acm.org/10.1145/1978942. 1979407.

67. J. Hullman, N. Diakopoulos, E. Momeni, and E. Adar. Content, context, and critique: Commenting on a data visualization blog. In *Proceedings of Conference on Computer Supported Cooperative Work (CSCW)*, Vancouver, 2015.

# Index

Milton Keynes UK
Ingram Content Group UK Ltd.
UKHW031128141024
449569UK00006B/351